EIGHT EUROCENTRIC HISTORIANS

EIGHT EUROCENTRIC HISTORIANS

J. M. Blaut

THE GUILFORD PRESS
New York / London

© 2000 J. M. Blaut
Published by The Guilford Press
A Division of Guilford Publications, Inc.
72 Spring Street, New York, NY 10012
www.guilford.com

Printed in the United States of America

This book is printed on acid-free paper.

Last digit is print number: 9 8 7 6 5 4 3 2 1

Library of Congress Cataloging-in-Publication Data

Blaut, James M. (James Morris)
  Eight Eurocentric historians / J.M. Blaut.
  p. cm.
  Includes bibliographical references and index.
  ISBN 1-57230-590-8 (cloth)—ISBN 1-57230-591-6 (pbk.)
  1. Historians—Europe. 2. Eurocentrism. 3. Historians—United States.
  4. Historiography. 5. Civilization, Western—Philosophy. I. Title.

D16.9 .B493 2000
940′.072—dc21

                                                    00-039339

Chapter 4 is a revision of an article with the same title that was published in *Antipode*
(Blaut, 1994); adapted by permission of Blackwell Publishers. Parts of Chapters 8 and 9
were incorporated in a paper, "Environmentalism and Eurocentrism," published in *The
Geographical Review* (Blaut, 1999); adapted by permission of the American Geographical
Society.

To Meca and Gini

Contents

Preface

This is the second of three volumes in the multivolume work titled "The Colonizer's Model of the World." The project as a whole is a critique of Eurocentrism in world history and historical geography. I try to demonstrate that our understanding of the human past will be much improved after we have sifted out and discarded those arguments and theories that falsely attribute historical superiority or priority to Europeans over all other peoples. I try to show that there is a systematic cultural–historical reason why we tend to hold on to these factually false beliefs. They fit within a world model that I label "Eurocentric diffusionism": the fundamental assumption that progress is somehow permanent and natural in the European part of the world but not elsewhere, and progress elsewhere is mainly the result of the diffusion of innovative ideas and products from Europe and Europeans. I argue that this model gained its validation and power from European colonialism, which seemed to confirm the superiority of Europeans over everyone else.

The first volume in this series, *The Colonizer's Model of the World: Geographical Diffusionism and Eurocentric History*, examined the nature and evolution of Eurocentric diffusionism; discussed beliefs about European superiority of mind, culture, and environment; and then sketched in the bare outlines of a theory that tries to explain the rise of Europe in a non-Eurocentric way. The present volume has a different but complementary purpose: to examine, and try to refute, eight distinct Eurocentric theories about world history that are important and influential in contemporary social thought. The third volume will put forward a non-Eurocentric historical model of the period from late medieval times to the nineteenth century and will also discuss a number of theoretical issues in the critique of Eurocentric history, among them the matters of Eurocentric Marxism, Euro-environmentalism, and Malthusianism.

Many people contributed in important ways to the writing of this book and to the development of the ideas (the good ones, not the bad ones) that it puts forward. Wilbur Zelinsky, who is one of the most important historical geographers of our time and a valued friend, assisted (so to speak) at the birth of this book. After seeing and commenting on a draft of Volume 1, he said to me, in essence: "Jim, you won't persuade the scholarly community that some of their most cherished beliefs about history are false simply by refuting these beliefs—you need to take on the historians themselves, the influential expounders of Eurocentric world history, and answer their specific arguments." That is what I try to do in the present volume. I have also had help and wise counsel from other colleagues and friends, including Janet Abu-Lughod, Raymond Brod, Andre Gunder Frank, José López, Kent Mathewson, Louis Proyect, Anselme Rémy, David Stea, Peter Taylor, Ben Wisner, and others too numerous to mention; but most importantly from my wife and principal adviser, América Sorrentini de Blaut. Peter Wissoker of The Guilford Press has helped in many ways, intellectual as well as editorial. Since this book continues an argument begun in the first volume of this series, I have had to return in a few places to the text of the earlier volume, paraphrasing a number of paragraphs and intoning, in many footnotes, "see Volume 1."

CHAPTER 1

Eurocentric History

B y my reckoning four kinds of Eurocentric theory have been advanced to explain the fact that Europe (or the West) grew richer and more powerful than all other societies. The four are:

1. *Religion*: Europeans (Christians) worship the True God and He guides them forward through history.
2. *Race*: White people have an inherited superiority over the people of other races.
3. *Environment*: The natural environment of Europe is superior to all others.
4. *Culture*: Europeans, long ago, invented a culture that is uniquely progressive and innovative.

These doctrines have been used in various combinations. Early in the nineteenth century the religious doctrine was prominent, but historians had no hesitation about invoking race, environment, and culture as God's instruments. Later, overtly religious explanations became unpopular, and Europe's superiority was then attributed mainly to race, culture, and environment. Now racism has been rejected, and Eurocentric history stands today on just the two legs: environment and culture. Europe, we are told, rose and conquered the world because its environment and its culture are superior: they caused Europe to develop faster and further than every other society.

All of this, I argue, is wrong: it is false history and bad geography. Europe's environment is *not* better than the environments of other places— not more fruitful, more comfortable, more suitable for communication

1

and trade, and the rest. Europe's culture did *not*, historically, have superior traits, traits that would lead to more rapid progress than that achieved by other societies: individual traits like inventiveness, innovativeness, ambitiousness, ethical behavior, etc.; collective traits like the family, the market, the city. The rise of Europe cannot be explained in this Eurocentric way.

I believe that the rise of Europe—that is, the surging of Europe past other civilizations in wealth and power—did not begin until 1492, and resulted not from any unique preexisting internal qualities but from Europe's location on the globe: Europe had immeasurably greater access to the riches of the New World than did any other Old World civilization.

This argument was advanced in the predecessor to this book, *The Colonizer's Model of the World: Geographical Diffusionism and Eurocentric History*, and I will summarize the argument in a later section of this chapter.[1] However, a word needs to be said on the subject now, by way of introduction. This theory has two parts, only the first of which is of primary concern in the present volume. First there is the argument that, prior to about 1500, Europe was roughly on a par with several other civilizations in terms of economic and technological development. Here I stand in opposition to the great majority of traditional historians. They differ among themselves about the explanation for the post-1500 rise of the West, but almost all of them agree that the explanation is mainly to be found in the Europe of pre-1500 times, namely, in its superior environment or its more advanced or more progressive culture, or (usually) both. Nearly all of the currently important arguments that support this uniqueness-of-Europe position are advanced by the eight historians whom we discuss in this volume. The second theory is an effort to explain the rise of Europe after (about) 1500 without recourse to the uniqueness-of-Europe position. I believe that the process can be explained in terms of this fundamental fact: Europe acquired incalculable riches from the Americas after 1492. This led to the rise to political power of the merchant–capitalist class and its allies, and in many other ways led, directly and indirectly, to the awakening of Europeans to the rest of the world and the transformation of Europe's society and economy. Here was an entire hemisphere, North and South America, six times the size of Europe itself, almost emptied of its population by the importation of Old World diseases during the sixteenth century, and immeasurably more accessible to Europeans than to any other civilization. This theory was discussed in Volume 1 and will be elaborated much more fully in Volume 3, but little will be said about it in the present work, which seeks to refute the arguments proclaiming the uniqueness of premodern Europe, and does so in a way that is rather novel and (I hope) fruitful.[2]

Eurocentric world history is more than a theory: it is a vast complex of beliefs, a world model, made up of countless statements of fact and explanatory theories. These are connected up in a very loose and incomplete network of arguments. Some of the arguments are tied to good evidence, some to poor evidence, some to no evidence at all (often they are inherited folk myths); some arguments are validated by others and some float free. The most highly structured arguments are the written narratives presented by individual historians, and these are what we will look at in this volume.

When a scholar writes about world history or some major part or aspect of it, he or she naturally wants to connect up the various statements of fact and theories so that they form a coherent and plausible argument. Typically, today, the narrative, if it is Eurocentric, is made up of a great number of strikingly diverse arguments: European superiority or priority in everything from climate and topography to demography, technology, state, family, and mentality. My task in this book is to look at narratives of this sort that are put forward by eight important Eurocentric historians.

A work that examines sequentially the writings of eight scholars who put forward Eurocentric views of world history will inevitably have a certain amount of duplication, since these scholars employ many of the same arguments; indeed, this commonality is one of the key generalizations to be made about the school as a whole in the concluding chapter. But the eight historians present differing forms and variants of these common arguments and theories (for instance, Mathusianism, Oriental Despotism, European "rationality"), and this calls for somewhat different responses and citations in each case. I will therefore have to deal with some of the formulations more than once. But repetition will be kept to the minimum, and mostly will be avoided by referring to prior discussions of each argument.

EUROCENTRIC DIFFUSIONISM

This book builds upon the arguments of Volume 1, and perhaps it will be useful to some readers if I very briefly review two of the principal arguments presented in the prior work: the nature and history of the world model that can be called "Eurocentric diffusionism" and the skeleton of a non-Eurocentric theory about colonialism and the rise of Europe. "Eurocentrism" is a term that was coined fairly recently to indicate a subspecies of ethnocentrism. The latter word in its most common usage means thinking or action or discourse that is "centered" on one ethnic community and falsely claims for that community a superiority, in worth or characteristics, over other communities or the rest of the world. I use the term "Eurocentrism," then, to indicate false claims by Europeans that their society or region is, or was in the past, or always has been and always will be, superior to other societies or regions.[3] The key word here is "false." It is not Eurocentric to prefer European music to other music, or European cuisine to other cuisine. It *is* Eurocentric to make the claim that Europeans are more inventive, innovative, progressive, noble, courageous, and so on, than every other group of people; or that Europe as a place has a more healthy, productive, stimulating environment than other places. It is not Eurocentric to extol "England's green and pleasant land," but it *is* Eurocentric to claim that this land is greener and more pleasant than all the other lands of the world.

Eurocentrism has been around since early medieval times, when Europe as a place began to take form in the minds of the community that inhabited it. The community in those times was generally thought of as Christendom, not Europe per se. Christendom was thought to be superior to all other societies because Christians worshipped the True God and therefore they, and their community, and their region on earth must naturally come under His protection. But medieval Christians knew that their civilization was not superior, in terms of wealth, technology, and most institutions, to their Islamic neighbors to the south and east. Europeans—Christians—were superior to everyone else in the most important way of all: they had the greater hope of being admitted to heaven.

Modern Eurocentrism really began in 1492. When Columbus returned from his first voyage to America, he described a people who were heathens, and who, he believed, could be conquered easily. Moreover, the conquest of their land would provide gold and other wealth to Europeans. It seemed clear that Europeans were superior to these Americans and would profit from this superiority. The conquest did indeed prove fairly rapid (mainly because the American populations were decimated by introduced Eastern Hemisphere disease), and the profits were indeed immense. Europeans could now, for the first time on a significantly large scale, make a clear distinction between themselves and a non-European people to whom they could really believe themselves to be superior. The Eurocentrism that thus emerged in the sixteenth century had two essential characteristics: superiority seemed to be confirmed by the success of colonialism; and superiority produced great profits.

Colonialism proved even more successful in later centuries, eventually reaching the level where Europeans were able to conquer and rule not only the Americas but also most of Asia and Africa. And European endeavors in all of these continents continued to be hugely profitable. So Eurocentric beliefs seemed to be continually confirmed as both true and useful, and they gradually evolved into the Eurocentric world-model of modern times. When this model was fully developed, in the nineteenth century, it comprehended a conception of the history and geography of the entire world. And it became the mirror in which Europeans came to see themselves and their own past.

Eurocentrism has a geography as well as a history. The geography, at a world scale, is Eurocentric diffusionism. Visualize a landscape on which there lie many separate and distinct societies. A new trait—say, a tool or an art style—is invented in one of these societies. Later, the trait is found to be present in another society, somewhere else on the landscape. The new trait may have diffused, that is, spread, from the first society to the second—or the second society may have invented the trait on its own. These two cases are called, respectively, diffusion and independent invention. Some scholars have a tendency to believe that most human beings, in most societies, are imitative, not inventive, and when a case like this comes up, with the new trait appearing in a second society (near or far), these scholars tend to assume or argue that this must be a case of diffusion, not independent invention, because—since people are supposedly rather uninventive—independent invention is much rarer than diffusion. These scholars are traditionally called "diffusionists," and a kind of scholarly war has been going on in European social science for more than a century between the "diffusionists" and the "independent-inventionists" (sometimes called "evolutionists").

Diffusion can occur at any scale. It may reflect the spread of something from one individual to another, or one society to another (as in the example above), or it may occur on an entirely world-wide canvas. Diffusionists usually (though not always) tend to explain culture change in terms of diffusion rather than independent invention at all of these scales. Many scholars in the past, and a few in the present, are called "extreme diffusionists"; they argue that all of culture began at one place on the face of the earth and spread out from there to the rest of the world. The original center from which diffusion occurred was always the region that can be called "Greater Europe." I use this term to refer to the continent of Europe plus (for ancient times only) the Europeans' self-proclaimed culture hearth, the "Bible Lands," and (for modern times only) outlying European-settled regions such as Anglo-America. Until fairly recently, Western scholars believed that essentially all of the important and progressive cultural advances since ancient times occurred somewhere in Greater Europe. In traditional European scholarship it was believed that Greater Europe naturally invents, innovates, progresses; the rest of the world remains stagnant and unchanging, or (like China) progresses only slowly and fitfully. Basically all of the history-making inventions and innovations were thought to have originated in Greater Europe, which supposedly invented agriculture, metallurgy, cities, states, social classes, democracy, science, most of the fine arts, and much more.

This model was not precisely diffusionist because most scholars (other than some of the extreme diffusionists) were aware that Europe did not have a significant impact on the rest of the world prior to 1492; the model was, in essence, a map of the world with two permanent regions, one of which, Europe, "the core," was always progressive, always superior; the other, non-Europe, "the periphery," was always backward. Since the periphery was considered marginal to social evolution, European historians tended to describe and explain both world history and European history with a kind of tunnel vision that may be termed "tunnel history." To explain any fact, event, or process, the historian would look only at prior facts, events, or processes within Europe itself, essentially ignoring the rest of the world. It was as though they looked back through a tunnel of time whose walls were the boundaries of Greater Europe; outside of these walls, everything was rockbound, timeless tradition.

In the study of the modern era, tunnel history fused with diffusionism. European history was still tunnel history. When discussing modern times, after 1492, European scholars tended to argue that progress since the sixteenth century for non-Europe has been a process of diffusion from Europe through the mechanism of colonialism. So now we have a model that retains the two world sectors, core and periphery, but sees a steady diffusion outward from the core, consisting of the fruits of European inventiveness, innovativeness, and progress, with a counter-diffusion into Europe of colonial wealth. This model is now the fully developed, classical, model of Eurocentric diffusionism, the model that dominated European scholarship until well into the twentieth century.

Classical Eurocentric diffusionism was grounded in five fundamental propositions:

1. Progressive cultural evolution in Greater Europe is self-generated, autonomous, natural, and more or less continuous.

2. Progressive evolution in Greater Europe results mainly from the action of a force or factor that is ultimately intellectual or spiritual; it is European "rationality" (inventiveness, innovativeness, ethical judgment, and so on), and is the primary source of European progress in technology, in social, economic, and political institutions, in science, art, and religion. This quality of superior rationality reflects either racial superiority, or a superiority of culture that originated in ancient or medieval times, or a superiority of Europe's natural environment.

3. Non-European regions and societies do not, in general, change as a result of their own internal causes; they change as a result of the diffusion of innovations coming directly or indirectly from the European sector.

4. The main form of interaction between Europe and non-Europe is the outward diffusion of progressive innovations (ideas, things, settlers—in aggregate, civilization) from Europe to non-Europe.

5. A natural consequence of this outward diffusion is the return flow, the counter-diffusion, from non-Europe into Europe, of wealth in the form of precious and nonprecious metals, plantation products, art objects, and other valuable things, a sort of partial repayment for Europe's gift of civilization.[4]

Classical Eurocentric diffusionism was the intellectual model associated with a period in which diffusion was taking place rapidly and profitably, namely, the era in which colonialism flourished and colonial rule seemed to be a permanent or at least long-lived condition. It was also an era in which Europeans knew relatively little about the history of non-European regions; and most of what they learned came from writers who were themselves close to or part of the diffusion process—missionaries, colonial administrators at home and abroad, wealthy travelers, and the like—and who, in most cases (though by no means in all cases) did not expect to find that the past of the colonized peoples had displayed progressiveness. This era ended, by stages, in the twentieth century: there was the shock to European power and self-image caused by the two world wars and the Great Depression; later, after 1945, there was decolonization. Classical Eurocentric diffusionism did not, however, end: it became transformed into modern Eurocentric diffusionism. This transformation resulted from profound changes both in society and in scholarship. These were delineated in Volume 1, and here I will discuss only one outcome: the present-day Eurocentric historiography that is our primary concern in the present volume.[5]

The critique of Eurocentric diffusionism began (roughly) in the 1930s with the work of a few historians from the European and colonial worlds.[6] After 1945 there was a rapid acceleration in research by historians, European and non-European, into the history of Asia, Africa, and Latin America. The new evidence (along with a change in attitudes) made it necessary to reject the older notion that non-Europe had been stagnant and "ahistorical" (the old Hegelian code-word) until the coming of the Europeans. But the essential doctrine, stripped of racism and colonial prejudices, remained hegemonic: non-Europe had not progressed at the same overall rate as Europe, although some regions (notably China) had been progressive in certain periods. Tunnel history was modified. Europe had indeed received important diffusions from non-Europe, although the priority of Greater Europe in most of the transformations (urbanization, industrial technology, democracy, etc.) remained largely unquestioned. And a somewhat new form of tunnel history in part replaced the old, an approach that coopted the idea of comparative history, supposedly looking at non-European as well as European history, but actually, in most cases, simply juxtaposing negative (often prejudiced) statements about non-European history with positive statements about European history. The proposition that counter-diffusion of wealth from non-Europe to Europe is natural was rejected: Eurocentric scholars now argued that the diffusion of wealth, both during the colonial period and more especially after decolonization, goes *from* Europe *to* non-Europe. Under this perspective, colonial development, international aid programs, and "globalization" diffuse resources to the non-European world, a world that had been originally sunk in poverty until Europeans came and brought modernity and "development."

This is Eurocentric diffusionism in its modern form. It has been subjected to strong and increasing criticism during the past two or three decades by a number of scholars who are trying to reformulate world history in a non-Eurocentric way.[7] This book continues that critique.

THE RISE OF EUROPE

The part of the Eurocentric-diffusionist model that has been most resistant to change is the theory of the rise of Europe. Most historians still maintain that this process was mainly self-generated—that it was mainly a result of the uniquely progressive culture of Europeans, especially their rationality; and that Europe's physical environment was somehow uniquely favorable in terms of productivity and thus is part of the explanation for Europe's rise and triumph.

In Volume 1 I presented a skeleton history of the European and non-European world from the fourteenth century through the seventeenth century, in order to show that the rise and triumph of Europe did not result from any prior actual or potential superiority over other civilizations, but resulted, rather, from the immense wealth that flowed into Europe from the Western Hemisphere and later from other colonized regions. What follows here is a very brief summary of that historical theory. Obviously, a summary merely lays out the propositions; the underlying arguments and evidence were presented in Volume 1 and will not be repeated here, although many of these arguments, and some of the evidence, will emerge in later chapters as I counter Eurocentric falsehoods.

I argue, first, that Europe and several non-European civilizations in the Eastern Hemisphere were progressing at roughly similar rates, and had attained roughly similar levels of economic and technological development, in the period just prior to A.D. 1500. The rural society of Europe was not dissimilar in nature to rural societies of India, China, and some other regions. All were basically landlord-and-peasant social systems. Peasant agriculture was the centerpiece of the economy, with comparable levels of commercialization and labor productivity in many areas. In some regions serfdom was or had been important; in others, tenant farmers had some degree of freedom under the landlords' rule; but Europe was by no means the only region in which serfdom dominated rural society during most of the Middle Ages. The estates or manors of Europe were not significantly different from those of many other regions: Europe's feudal estates were not (as Max Weber famously argued) closer to true private property than were the estates of many other regions in the hemisphere.

The levels of urbanization characteristic of Europe were also characteristic of many other civilizations in the fifteenth century; most importantly, the maritime–mercantile trading cities of Europe had counterparts along the coasts of the Indian Ocean and the western Pacific, and business institutions (like banking and accounting) were also comparably sophisticated in European and non-European trading cities. In the fifteenth century, Europe's maritime trading centers were pretty much at the same level of development as many other such centers. All of them were engaged in long-distance oceanic trade, and many were trying to enlarge that trade; exploration was part of this process, in the Indian Ocean and the Pacific as well as the Atlantic. Technology, in agriculture, manufacturing, civil engineering, and other spheres, was not more highly developed in Europe than in China and probably in the other great civilizations.[8] European minds were not more rational than the minds that dwelt in other regions. Indeed, the various civilizations were profoundly different in many aspects of culture, such as religion, but these differences did not devolve into differences in the levels and rates of development of the ecological dimensions of culture, notably technology and the economy.[9]

If it is accepted that Europe in 1500 was neither more advanced nor more progressive than several other regions, how can we explain the fact that Europe *after* 1500 began to develop more rapidly than the rest of the hemisphere and eventually came to dominate the world, experience an industrial revolution, and, in the broadest terms, "rise"? My answer to this question is that the immense, almost incalculable wealth that Europeans obtained in colonial adventures, from precious metals, plantation agriculture, and unequal exchange, provided the development-oriented classes in part of Europe—merchants, traders, manufacturers, landlords investing in agricultural improvements and trade, and other commerce-minded groups—with the political power to wrest control of their societies from the feudal landlord class and its allies. Europeans very quickly got hold of immense quantities of gold and silver. For an example: from 1500 to 1800 eighty-five percent of the world's silver and seventy percent of its gold came from the Americas.[10] Europeans acquired abundant fertile land, as a result mainly of depopulation, and by the end of the sixteenth century Europeans had already begun to reap great profits from plantations. (In 1600 the value of Brazilian sugar was double the value of all of England's exports to all of the world.[11]) The Europeans, one might say, got rich quick.

In the sixteenth century most of the wealth came directly or indirectly from the Americas in the form of precious metals, these serving both to dissolve (so to speak) the old social forms in Europe and to permit lucrative trade with Asia and Africa, hence generating even more wealth for this class-community. I believe that this community, which I call "protocapitalist," essentially bought out the old feudal class (that is, drew accumulation-minded landlords into its financial fold) and took control of the society, first in Holland and then (with the "Glorious Revolution" of 1688) in England, with parts of other countries also experiencing the same profoundly important transformation. When these societies came to be dominated by capitalism, there began several other transformations dependent on political power (and wealth), such as the forced creation of a nonagricultural working class, the use of state funds for further colonial enterprises, and much more. In Volume 1 I elaborated these arguments and carried them forward to the eighteenth century, arguing that the rise of Europe from 1500 to 1800 was mainly fueled by a process external to Europe itself: by colonialism. Great progress occurred in Europe during this period, and the immediate causes are mostly to be found within Europe itself. But the underlying cause was colonialism: the constant flow of wealth that was yielded by formal and informal colonialism, the life opportunities created by colonialism, the new ways of thinking and new inventions that were stimulated by colonialism, along with inflowing diffusions of ideas and techniques from other continents. All of this, in my view, is the basic underlying dynamic of the rise of Europe and European capitalism.

So the two basic arguments are: (1) Europe was not more advanced or more progressive than other civilizations in 1500; and (2) colonialism explains, at the most fundamental level, the rise of Europe after 1500. But there is still a third important question: how did it come to pass that Europeans, and not Asians or Africans, acquired colonial empires and the wealth therefrom? We should bear in mind that the Atlantic explorations of the Iberians had counterparts in the explorations by non-Europeans in the Indian Ocean and the Pacific during the same period. Many of the mercantile–maritime centers of the hemisphere, in all coastal regions, were extending their radii of trade and travel. The Europeans had, as it turned out, one crucial advantage. During the sixteenth century, the source of colonial wealth was America; and America as a whole, but more importantly the parts of America where gold and silver were widely used, was vastly more accessible to European than to other maritime centers. The distance from the Canaries to the West Indies was one-third the distance from China to Mexico (Acapulco), while the distance from East African ports to the West Indies was almost as great as that from China to Mexico. Moreover, the Atlantic wind systems consist of easterlies in the tropical and subtropical latitudes and westerlies in higher latitudes; this wind system was familiar to European sailors from voyages to the Canaries, Madeira, and the Azores, and it was known that one could sail westward outbound, in the trade wind belt, and return eastward with the midlatitude westerlies (hence, an extension of the known route westward to—as they thought—Japan seemed eminently feasible). By contrast, wind circulation in the north Pacific is much less favorable and reliable.

I argue, therefore, that the Americas were reached first from the mercantile–maritime centers of Europe, rather than from those of other advanced mercantile centers, as a function of location or, more precisely, accessibility. Now America had been isolated from the other hemisphere for millennia, and Americans succumbed rapidly and massively to Old World diseases; in Mexico, for instance, at least four-fifths of the population died during the sixteenth century.[12] Thus, the conquest of areas that could be looted and mined for gold and silver proceeded very rapidly.[13] The accumulated wealth permitted Europeans to quickly attain a level of maritime power that made it impossible for other civilizations to force their way into the American treasure trove. In sum: Europeans acquired the wealth from colonialism because of their location on the globe, not because they were somehow uniquely advanced, or progressive, or "venturesome."

THE PROJECT

The present volume is the second of a three-volume work, *The Colonizer's Model of the World*. The project as a whole has one basic purpose: to uncover, criticize, and refute as much of the corpus of Eurocentric theories and truisms in world history and social thought as I can manage to do. (Recall that I label an idea as Eurocentric only if it *falsely* attributes some uniquely positive quality to Europe and Europeans, past or present.) Volume 1, subtitled *Geographical Diffusionism and Eurocentric History*, dealt with three parts of the overall problem. First, it analyzed Eurocentrism as a world model, and analyzed more concretely the portion of this model that influences world history, namely, Eurocentric diffusionism. It tried to locate this doctrine within the history of Western ideas and tried to show that the social force that has driven it throughout modern history has been colonialism; that this doctrine is indeed the colonizer's model of the world. Next in Volume 1 came a systematic critique of the myths that claim Europe's historical superiority in matters of race, environment, mind, and society. Finally, I sketched in, very briefly, what I believe is a non-Eurocentric model of the history of the world from late medieval times to about 1750, trying to show (1) that Europe had no actual or potential advantage over the rest of the world prior to the conquest of America and (2) that colonialism, beginning with the Conquest, was the basic force that led Europe to rise and modernize.

Volume 2, the present work, deals with Eurocentric history as a body of theories put forward by eight notable historians. Volume 3, subtitled *Decolonizing the Past*, will have two purposes. The first of its two parts is a series of essays on aspects of Eurocentric diffusionism in social thought. The second part is a fully developed model of world history from late medieval times to the nineteenth century, with considerable attention to the role played by colonialism in the Industrial Revolution. The three-volume project as a whole is intended to be a contribution to the decolonization of world history and social science.

THE HISTORIANS

Chapters 2 through 9 are critical studies of the work of eight Eurocentric historians. The first one, Max Weber, was chosen for analysis because his ideas, now more than seven decades after his death, still underlie much of contemporary Eurocentric historiography and have at least some influence on almost all of it. The other seven scholars are moderns. I chose them for study on the basis of a combination of criteria.

Five of the historians (Eric Jones, Michael Mann, John Hall, David Landes, and Jared Diamomd) present entire, global, world-historical arguments, endeavoring to show that Europe had superiority over all other world regions throughout millennia of history for a multitude of reasons. These five scholars may be, right now, the most important, or at any rate most widely read (in the Anglophone world), of truly Eurocentric world historians. Before treating their work we will discuss two other historians, the late Lynn White, Jr. and Robert Brenner, who present arguments on a more modest issue, yet one that is absolutely crucial to Eurocentric history: why Europe rose out of medieval backwardness and became richer and stronger than other civilizations (the problem sometimes known as the "transition" to capitalism and modernity). White's and Brenner's Eurocentric theories about the unique rise of Europe in medieval and early modern times are highly influential. Both White and Brenner posit an agricultural revolution in later-medieval northwestern Europe, White seeing this from a conservative perspective, Brenner from a Marxist one, and both view the "transition" as having taken place mainly in agriculture and entirely in northwestern Europe. I chose these eight scholars mainly for the foregoing reasons, but also because, as a group, they present what seems to me to be almost the full array of currently popular theories about Europe's superiority or priority in history.

The critical analyses appear in, roughly, the chronological order of the principal writings of each historian. Max Weber (Chapter 2) is discussed first, partly because he is the earliest of the historians whose views we will discuss, but mainly because of his immensely important influence on each of the other seven, and indeed on modern historical scholarship as a whole. Weber was a German sociologist and historian whose work extended from the later nineteenth century into the 1920s. He is rightly considered to have been one of the greatest social scientists of his time. He laid down much of the foundation for modern sociology. His historical works are exceedingly important.[14] But our concern here is with Weber's many writings in which he put forth and defended a viewpoint on what he considered to have been the unique progressiveness of Europe through millennia of history, and rooted that supposed uniqueness mainly in "European rationality." This thesis has been extremely influential in modern historiography. It forms part of the grounding for most present-day Eurocentric formulations of world history; and it underlies, in part, what may be the most significant Eurocentric doctrine of recent social science: "modernization theory," the diffusionist view that the non-European world develops by receiving modernizing ideas and material things from the European world. Since Weber is the fountainhead of so much contemporary Eurocentric thought, it seemed important to analyze his view of world history along with those of seven contemporary scholars.

Lynn White, Jr. (Chapter 3), was a respected medievalist historian, a specialist in the history of technology. His 1962 book *Medieval Technology and Social Change* has had substantial influence on later Eurocentric historiography. It presents a technologically deterministic theory of Europe's medieval rise, arguing that Western Europe was uniquely inventive in medieval times; that this society accomplished, uniquely, an agricultural revolution; and that this putative medieval agricultural revolution is a large part of the explanation for the supposed rise of Europe before the end of that period. White's viewpoint is important for us partly because it is a classic statement of the technologically determinist perspective in Eurocentric history. Partly, also, he describes (in other works) his belief that Europe's technological uniqueness in the Middle Ages derives from special qualities in the Judeo-Christian tradition and the Christian religion.

Robert Brenner (Chapter 4) is a Eurocentric Marxist historian. He warrants our attention partly because of the influence that his work has had on mainstream as well as Marxist historiography, partly because his arguments are rather original, focusing on late-medieval rural class struggle, and partly because he can serve as a kind of token for the very active modern school of Eurocentric Marxist historians.

The other five historians put forward comprehensive theories about Europe's historical superiority or priority. ("Superiority" is used in this book to mean greater possession of cultural or environmental qualities that favor development. In some contexts I use the word "priority" to in-

dicate the more modest view that Europe merely arrived at a given historical point at an earlier date than did other societies.)

Eric L. Jones (Chapter 5) is an economic historian whose work on
early-modern European history, especially agricultural history, is solidly
grounded and highly respected. However, his 1981 book *The European
Miracle: Environments, Economies, and Geopolitics in the History of Europe
and Asia* is an extreme example of Eurocentric world history. It set in motion several new trends in Eurocentric historiography and popularized
the expression "the European miracle." In this work he advanced what is
certainly the most influential modern argument for Europe's superiority
throughout history; the title phrase "the European miracle" has almost
become the signature of Eurocentric historiography today; and I therefore
devote more attention to this historian than to the others.

Michael Mann (Chapter 6) is a historical sociologist, much influenced by Max Weber and Ernest Gellner, whose widely discussed 1986
book *The Sources of Social Power: Vol. 1. A History of Power from the Beginning to A.D. 1760* presents a view of world history that is exceedingly
Eurocentric. Mann proposes to offer a theory of history that grounds itself
in the teleological idea that the core of civilization ("social power")
moves steadily westward. This supposed geographical march of history is
the most consequential part of his argument.[15]

John A. Hall (Chapter 7), a political scientist and historical sociologist, is, like Mann, strongly influenced by Weber and Gellner. His 1985
book *Powers and Liberties: The Causes and Consequences of the Rise of the
West* and later essays put forward a Eurocentric theory of world history focusing on political and state-forming processes. He claims that European
culture alone possessed the qualities that would permit political modernization.

Jared Diamond (Chapter 8) is a bioecologist whose 1997 Pulitzer
Prize-winning book, *Guns, Germs, and Steel: The Fates of Human Societies*, presents the most extreme example of environmental determinism
in the service of Eurocentric world history that we have seen in the past
half-century. In it he argues that the natural environment alone explains
the superiority of midlatitude Eurasia over all other world regions, and
Europe's rise above other Eurasian regions is a product of environment
helped along by culture. This book seems to be generating a revival of the
extreme environmental determinism that was popular in nineteenth and
early twentieth-century scholarly thought but was rejected forcefully by
geographers before the middle of the twentieth century.

David Landes (Chapter 9) is an economic historian who has made
important contributions, mainly in the area of the history of European

technology. His 1998 book *The Wealth and Poverty of Nations: Why Some Are So Rich and Some So Poor* was enthusiastically reviewed in *The New York Times*, *The Wall Street Journal*, and the *Washington Post*, in what can only be described as broad approval of Eurocentric world history on the part of opinion-formers in American society. This book is remarkable for the great number and variety of traditional arguments that it makes on behalf of Landes's claim that Europe has been superior to the other cultures of the world in environment, economy, technology, politics, society, and mentality since ancient times. The book is, just at the present moment, the most widely discussed example of Eurocentric world history.

These eight historians are not, of course, a representative sample of Eurocentric historiography. The seven contemporary scholars, however, seem to me to provide almost the entire spectrum of Eurocentric arguments that are being widely used today to explain the rise and triumph of Europe. In Chapter 10 I list thirty such arguments. Also, in spite of differences among the eight scholars, they seem to use what I think is the standard model of present-day Eurocentric world history. In Chapter 11 I sketch in this model.

Finally we come to the question whether the Eurocentrism displayed by the eight historians discussed here is somehow typical of modern Western historiography. There is a strong thrust in modern historiography, both in scholarly works and in textbooks, toward eliminating old Eurocentric beliefs. To some extent this reflects new knowledge about the non-European world, which no longer is dismissed by (most) scholars as somehow stagnant, traditional, and the like: the conventional view in Max Weber's time. To some extent this reflects the impact on scholarship as a whole of the civil rights movement and later trends toward cultural fairness. No longer do we find overt racism, or for that matter overt religious bias, in mainstream world history textbooks.[16] And there have been several forceful critiques of Eurocentric historiography—the most notable one being Edward Said's *Orientalism* (1978)—during the past three decades. So the answer I would give to the question posed above is: no, the Eurocentrism displayed by the eight historians discussed in this book seems *not* to be typical of modern historiography. But readers of this book will find (or know already) that a lot of Eurocentrism is still in evidence: *some* of the doctrines expressed by the historians discussed here are still accepted and professed by *most* historians. Among these doctrines are erroneous views about the European environment; erroneous comparisons between European and non-European history in such areas as urbanization, social class, technology, family, and so on; and a still widely held belief in the unique "rationality" of Europeans. If these erroneous beliefs were no longer current, I would not have written this book.

NOTES

1. The historical argument was put forward in Chapters 3 and 4 of Volume 1. The argument will be developed more fully in Volume 3 (subtitled *Decolonizing the Past*). "Volume 1" did not appear in the title of *The Colonizer's Model of the World: Geographical Diffusionism and Eurocentric History* (1993a) when it was published.

2. Two other non-Eurocentric models of the origins of Europe's rise must be mentioned. Janet Abu-Lughod, in part following William McNeill, suggests that the Black Death of the fourteenth century did much greater damage to the Asian economies than to the European economy, giving the latter, which had previously been no more developed than Asia, a great initial advantage. Andre Gunder Frank agrees with me (see Frank, 1992) that the silver and gold from the Americas initiated the rise of Europe, but he argues that Europe did not "rise" substantially until the eighteenth century. See Abu-Lughod, *Before European Hegemony: The World System A.D. 1250–1350* (1989); McNeill, *Plagues and Peoples* (1976); Frank, *ReORIENT* (1998). My emphasis on colonial accumulation as the prime cause of Europe's "rise" has been prefigured by earlier formulations, though none of these, as far as I can tell, rejected the assumption of at least some preexisting European uniqueness as part of the causal theory. See, for instance, R. H. Tawney, "Introduction" to Weber's *The Protestant Ethic and the Spirit of Capitalism* (1958); W. P. Webb, *The Great Frontier* (1951). Comments by Marx and Engels on the importance of the discovery of America are often cited, but it is clear that Marx and Engels viewed internal European forces—mainly class struggle—as more important than external capital accumulation. See Marx and Engels, *The German ideology* (1976).

3. "Eurocentrism" is also commonly used to mean simply an inordinate amount of attention—inordinate salience—given to Europe. Of course, it is used in still other ways as well.

4. A sixth proposition might be added. Because non-Europe is primitive, with archaic characteristics that Europe possessed in ancient times, there is another sort of counter-diffusion, also quite natural, consisting of evil, dangerous, and atavistic things like black magic, walking mummies, plagues, and vampires; see Volume 1.

5. I do not want to leave the impression that Western historians in the present period are prejudiced against non-European peoples, or that the corpus of writings on world history is to be dismissed as Eurocentric. See Chapters 10 and 11.

6. Among the most important were A. Appadorai, R. P. Dutt, C. L. R. James, J. C. Van Leur, and E. Williams. The work of James and Williams is briefly discussed in Volume 1. See Appadorai, *Economic Conditions in Southern India (1000–1500 A.D.)* (1936); James, *The Black Jacobins* (1937); Dutt, *The Problem of India* (1943); Williams, *Capitalism and Slavery* (1944); Van Leur, *Indonesian Trade and Society* (1955).

7. See, for instance, Abu-Lughod, *Before European Hegemony*; Amin, *Eurocentrism* (1989); Hodgson, *Rethinking World History* (1993); Frank, *ReORIENT* (1998) and his earlier *World Accumulation, 1492-1789* (1978); Said, *Orientalism* (1979); Wolf, *Europe and the Peoples Without History* (1982).

8. Needham, *Science and Civilization in China (1954-)*; al Hassan and Hill, *Islamic Technology* (1986); Flynn and Giraldez, *Metals and Monies in an Emerging Global Economy* (1997); Pomeranz, *The Great Divergence: China, Europe, and the Making of the Modern World Economy* (2000); Frank, *ReORIENT*.

9. This position partly rests in Julian Steward's concept of multilinear evolution. See his *Theory of Culture Change* (1955). Regardless of cultural differences, people in all societies work to improve the conditions that preserve and enhance life. The part of cul-

ture that is most directly involved in this effort, the ecological part, tends to be evolutionary over the long span: different paths are employed by different cultures, but they are parallel in the sense of striving to accomplish ecological goals.

10. Frank, ReORIENT; Barrett, "World Bullion Flows, 1450–1800" (1991); Flynn and Giraldez, Metals and Monies.

11. Simonsen, História econômica do Brasil, 1500-1820 (1944); Minchinton, The Growth of English Overseas Trade (1969).

12. Cook and Borah, Essays in Population History (1979); Crosby, The Columbian Exchange (1972); Blaut, The Colonizer's Model of the World: Geographical Diffusionism and Eurocentric History (1993) (cited hereafter as "Volume 1").

13. Columbus set sail across the Atlantic from the Canary Islands. North of Acapulco on the Pacific coast of America, and south of the West Indies on the Atlantic coast, precious metals were not widely used before 1492.

14. A number of biographies of Max Weber and studies of his works and their significance are available in English. See, for instance, Freund, The Sociology of Max Weber (1968); Gerth and Mills, "Introduction: The Man and His Work," in From Max Weber: Essays in Sociology (1946), pp. 3–76; K. Löwith, Max Weber and Karl Marx (1982); Marianne Weber, Max Weber: A Biography (1975).

15. A later work, The Sources of Social Power: Vol. II. The Rise of Classes and Nation-States, 1760–1914 (1993), is a much more important contribution to social history; however, since it deals only with modern Europe (in spite of its implicit claim to universality: "the rise of classes and nation-states"), it is of marginal concern to the present book and will be discussed only briefly.

16. Some nineteenth-century world history texts are discussed in Chapter 1 of Volume 1.

CHAPTER 2

Max Weber:
Western Rationality

W eber was one of the greatest European scholars of the late nineteenth- and early twentieth-century period, one of the founders of scientific sociology as well as a major historian, and one of the most effective intellectual defenders of and apologists for the conservative, bureaucratic society of pre-World War Germany. Weber analyzed that society, and more generally Western capitalist society, in meticulous detail and with great insight, although there were important limits to his insight. He did not really see the contradictions in that society—it is ironic that he died in the midst of the collapse of Germany just after the World War—and he held to the typical conceit of his time and place, and class, in thinking that contemporary European capitalism is the culmination of social evolution—if not the end product of evolution ,at least its highest achievement thus far. (He had doubts about further ascent in the future.)

RATIONALITY

Broadly, Weber saw social evolution as essentially an intellectual progression, an ascent of human "rationality," meaning intellect and ethics, from ancient to modern society. At each stage of that evolution people invented a range of new social institutions, like higher forms of the state, the legal system, the bureaucracy, the city, and so forth, and these, too, marched forward through history not as basic causes—the causes lie at the level of individuals, their minds and their actions—but as products of "rational" thought. Unsurprisingly, Weber concluded that modern European society is not only the most rational of all societies but also a product of conscious human choice, of voluntarism. The German capitalist accumulates profit because he chooses to do so as his duty, his "calling." Likewise the worker, who works not because of need but because it is his "calling."[1] There is no better way to defend a social system than to claim that it reflects the wishes, the free choices, of its members. It is not difficult to understand why Weber is today viewed with reverence by many of the more conservative Western social scientists. Yet, withal, he was a great scholar.

Weber's history was tunnel history. The march toward ever more rational society took place in Europe, among Europeans. Outside of the European tunnel of time, all societies were traditional and, in varying degrees, irrational. (How so we will see in a moment.) Why have Europeans at all times been more rational than non-Europeans? Weber, although painstaking in his comparison between the "rationality" of Europeans and the irrationality of other societies throughout history, says rather little as to why, in the first instance, Europeans displayed this rationality. He notes several "factors," one of which is race.

Like most European scholars of his time, Weber was a racist. Europeans are genetically superior to non-Europeans. But race is merely one of many "factors," and social science does not as yet, according to Weber, know how to measure the racial "factor."[2] So Weber approaches race in a cautious way. He believes that Africans are plainly inferior. This is shown by the occasional revealing comment. (Negroes are "unsuitable for factory work and the operation of machines; they have not seldom sunk into a cataleptic sleep. Here is one case in economic history where tangible racial distinctions are apparent."[3]) It is shown more clearly by Weber's almost absolute unconcern, in his discussions of social evolution and the comparison of civilizations, for Africa and Africans (other than ancient Egyptians, whom he did not connect to Africa in cultural terms). The same holds true for Native Americans. (Attached to the comment about "Negroes," quoted above, is a parallel comment about Native Americans' "unsuitability for plantation labor."[4] And they, like Africans, do not figure in Weber's evolutionary and comparative theorizing.) As to Asians, Weber's views are, however, cautious. Weber, for instance, concedes "the possibility that many of the Chinese traits which are considered innate may be the products of purely historical and cultural influences." Among these "traits" of the Chinese are: "striking lack of 'nerves'," "strong attachment to the habitual," tolerance for "monotony," "slowness in reacting to unusual stimuli, especially in the intellectual sphere," "horror of all unknown . . . things," "good-natured credulity," "lack of genuine sympathy and warmth," "absolute docility," "incomparable dishonesty," and "distrust . . . for one another" (which "stands in sharp contrast to the trust and honesty of the faithful brethren in the Puritan sects"—in Europe).[5]

As to race in general, Weber is also cautious. At the end of a long and important statement (which we will discuss in a moment) declaring how much more rational Europeans are than anyone else, Weber says that "it would be natural to suspect that the most important reason" for the difference "lay in differences of heredity," and admits that he, Weber, is inclined to think the importance of biological heredity is "very great" but the study of genetic "influences" is still not sufficiently advanced to give many answers, and we should emphasize the study of social and historical factors.[6] This is clearly a renunciation of the extreme racism that was popular in Weber's intellectual circles, but it does not alter the fact that Weber saw race as one primordial, or presociological, factor explaining the greatness of the Europeans. Moderate racism is in its own way an extreme position when it is inserted at the root of an argument for differences between human societies. Race becomes a kind of initial shove, starting the process of historical differentiation, and perhaps also a subtle influence that is at work everywhere and at all times giving just an iota of greater rationality to the ideas and decisions of Europeans compared with those made by non-Europeans.

ORIENTAL DESPOTISM

The second factor discussed by Weber is an old and familiar argument about the supposed influence of the physical environment on human affairs: the theory of "Oriental despotism."[7] Centuries before Weber's time it was widely believed that the great civilizations of Asia, and of Egypt, have acquired distinctive characteristics resulting from the fact that they occupy dry environments and depend upon irrigation agriculture. The old theory, which Weber simply adopts, argues as follows: In civilizations that depend upon irrigation the state must be "despotic"—undemocratic—because it must organize the process of water distribution and waterwork maintenance; must force the populace to work on these waterworks and to accept the necessary decisions regarding allocation of water. Europeans referred to this type of society as "Oriental despotism" and contrasted it with the supposedly freer kind of society that may emerge in environments where rain falls on every citizen's farm and so no outside authority is needed to manage the supply of water to the farm. This old theory had been used since the seventeenth century in various ideological contexts relating to non-Europeans. Asians were unprogressive because their societies were "Oriental despotisms." Europeans were free and therefore progressive, innovative, and so on. Europe was a free society in part because of the supposed independence of each peasant in decision-making, since his rain fell only on his farm. This theory has been woven into many modern arguments about the supposedly democratic character of European society throughout history—and back to the Neolithic age—as against the supposedly undemocratic society of the "Oriental despotisms."[8]

In Weber the argument is made in several places, most notably in his discussion of the reasons why the Chinese state did not evolve rationally. It was because of the classic syndrome of Oriental despotism. Actually, early China did not depend that much on irrigation. Weber had some awareness that this had been the case in North China, so he contrived a special argument—not a valid one—to the effect that the importance of canals for transportation, rather than irrigation, produced the same political effect.[9] In Europe the state evolved rationally:

> The crucial factor which made Near Eastern development so different [from Greek development] was the need for irrigation systems, as a result of which the cities were closely connected with building canals and constant regulation of waters and rivers, all of which demanded the existence of a unified bureaucracy. There was an irreversible character to this development, and with it went subjugation of the individual. . . . On the other hand in Greece . . . the position of the monarchs declined . . . and so began a development which ended . . . with an army recruited from yeoman farmers who provided their own arms. Political power necessarily passed to this class, and therewith started to emerge that purely secular civilization which characterized Greek society and caused capitalist development in Greece to differ from that in the Near East.[10]

The theory of "Oriental despotism" has no validity anywhere. In part it is bad geography: many Asian civilizations did not depend on irrigation; among those that did, rain-fed farming was usually found as well (Egypt being an exception); and irrigation came to be used in most parts of Asia not because the land was too dry to support agriculture but because irrigation *intensified* food production—was thus a *cultural* fact, therefore an *effect* of the social system, not a cause.[11] And in part the theory is merely (in its old forms) prejudice: Europeans viewed Asian society as inferior, as "unfree," and would have found some other rationalization for this belief had the theory of irrigation as the cause of "Oriental despotism" not been available.

Weber also alludes, glancingly, to other primordial causes of European uniqueness. He believes, for instance, that trading cities are "freer" than inland ones, and notes the pattern of trading cities in ancient Mediterranean Europe as compared with inland cities of the Asian empires.[12] (But the Phoenicians and other non-Europeans also had trading cities, as did the communities around the Indian Ocean and the China Seas.) The fact that Greek cities were, in this way, "free" led to the development of various modern European institutions, including democracy.

But the truly interesting thing about Weber's treatment of the historical development of European society and the comparison between European "rationality" and the irrationality and traditionalism of others is, surprisingly, the *lack* of a systematic attempt to explain the roots of the matter. The few references to race and environment are scattered throughout a massive output of scholarly works about European history and society. This lacuna can be explained in various complementary ways. One would be to credit Weber's good scholarly sense of caution in dealing with matters of causation where there is little evidence—matters of very ancient history, human genetics, and the like. This explanation must be right, as evidence everything we know about Weber's careful, thoughtful scholarship and also his unwillingness to accept the exaggerated racist theories, and the anti-Semitism, and environmental determinism, that were commonplace in the German scholarship of his time. But this cannot be the complete reason for Weber's apparent unwillingness to devote attention to basic causality. In addition, Weber adopted a view of society in which ideas and values, and the evolution of ideas and values, are treated as prime causes of social processes, social structures, and social change.[13] The most crucial ideas, for him, were religious. Weber's most widely known essay attempting to explain social history, *The Protestant Ethic and the Spirit of Capitalism*, establishes a causal model in which religious beliefs condition the ideas people hold about the world and the values they accept about proper conduct in the world; these become patterns of behavior or action, which, in turn, transform the economic and other institutions of society. If explanatory emphasis is to be given in this way to ideas, values, and religion, then any ultimate, primordial explanation would have to be one that explains religion and the human psyche. But Weber was philosophically an Idealist (of one of the neo-Kantian schools), and not prone to finding material explanations for psychological and spiritual phenomena. Weber may or may not have been religious (there is some question about this matter), but he was not prepared to invoke God or the human soul as ultimate causes of social fact. So explanations would remain incomplete.

But it seems probable that the most fundamental reason why Weber rather avoided discussing the most basic causes in social evolution—or "universal history," as he sometimes called it—is something rather different. In Weber's time, and before, European historians did not often doubt that there is some ultimate and unquestionable principle underlying what I have called in this book "tunnel history." This principle merely asserts that there *is* a reason why European history advances and why it advances according to internal principles, owing nothing important to the non-European world (Bible Lands aside), and why the non-European world does *not* evolve—except under the influence of diffusions from Europe. In those times you could *assume* that European history would move forward in the normal course. Rarely, therefore, did historians take on the task of explicating fully what the underlying principle really is. For some of them it was consciously thought of as the work of God. For others, a naturalistic cause was assumed to be present, and it was of some interest to inquire into that cause but it was not essential. Some, of course, rode hobbyhorses, for instance trying to uphold the doctrine of environmental determinism by showing how the environment truly is the underlying principle that explains European social evolution.[14]

In the case of Weber, we can only infer that he was prepared to assume that this causal principle was at work, that it somehow led to a greater rationality among Europeans than non-Europeans and led Europeans to, in essence, invent new traits and institutions by way of advancing historically. He gives abundant description of the way this process works throughout all of Western history, from the ancient Near East on forward. He gives meticulous descriptions and brilliant analyses of the social institutions created thereby, the legal systems, theologies, ethical systems, governmental forms, urban forms, land-ownership forms, and many more. So the racist and environmentalistic comments sprinkled here and there in his works are perhaps largely beside the point. For Weber, the European mind is the source of social evolution, and one need not try to get behind it to something more fundamental.

"ONLY IN THE WEST . . . "

Europe's superiority over everyone else is quite absolute. The matter is stated baldly in the very beginning of Weber's famous essay *The Protestant Ethic and the Spirit of Capitalism*:

[One who is] a product of modern European civilization, studying any problem of universal history, is bound to ask himself to what combination of circumstances the fact should be attributed that in Western civilization, and in Western civilization only, cultural phenomena have appeared which (as we like to think) lie in a line of development having *universal* significance and value. Only in the West does science exist at a stage of development which we recognize today as valid. [The] full development of a systematic theology must be credited to Christianity . . . since there were only fragments in Islam and in a few Indian sects. . . . Indian geometry had no rational proof; that was another product of the Greek intellect, also the creator of mechanics and physics. . . . The highly developed historical scholarship of China did not have the method of Thucydides. [All] Indian political thought was lacking in . . . rational concepts. [Rational] harmonious music, both counterpoint and harmony . . . our orchestra . . . our sonatas, symphonies, operas . . . all these things are known only in the Occident. . . . In architecture . . . the rational use of the Gothic vault . . . does not occur elsewhere. [The] Orient lacked . . . that type of classic rationalization of all art . . . which the Renaissance created for us. There was printing in China. But a printed literature . . . and, above all, the Press and periodicals, have appeared only in the Occident. [The] feudal state . . . has only been known to our culture. . . . In fact the State itself . . . is known [in the full sense] only in the Occident.And the same is true of the most fateful force in our modern life, capitalism. [The] concept of the citizen has not existed outside the Occident.[15]

This (and much more that I have not quoted) establishes the problematic for Weber's sociology of religion. The problem is to explain the uniqueness of the West. Weber goes on to say that this requires, most crucially, an explanation of the unique rise of capitalism in the West. And this, in its turn, calls for an analysis of the sociology of religion, or more precisely the sociological basis for the "economic ethics" of the world's religions. This having been stated, Weber goes on to the comment quoted previously about how it would be "natural to suspect that the most important reason" for the uniqueness of the West's rationality lies in "differences of heredity."[16] End of introductory essay. On to the substance of Weber's classic work, *The Protestant Ethic and the Spirit of Capitalism*.

Before we say anything more about Weber, we have to respond to his grand indictment of all non-Occidental societies. He is fundamentally wrong, on all counts, for three principal reasons. First, he is comparing twentieth-century Europe, with its science, mathematics, orchestras, and the rest, with ancient non-European civilizations and with contemporary civilizations which at this historic moment are crushed under colonial rule. This comparison is unfair. Second, he is factually wrong about many matters. European science, mathematics, and technology were in no way higher than Chinese and Indian science prior to early-modern times. After the rise of Europe, and particularly after the industrial revolution, you can expect both a flowering of science and an awesome increase in the scale and opulence of all other accomplishments—for instance, huge orchestras. But if comparisons are made for the period before 1492, when many of the world's civilizations were truly medieval, then Europe does not stand out. Not in science, not in art, not in law, not in the development of capitalism. And third, Weber was just projecting the rather standard prejudice of the European bourgeois gentleman of his time in his negative judgment about the art and culture of non-Europeans. He belittles their theology. Their music is not "harmonious." He does not appreciate or understand their architecture. Their art is not "rational." Part of this prejudice of course is lack of knowledge. (For an obvious example: Weber's bland assertion, just quoted, that "printed literature . . . and, above all, the Press and periodicals, have appeared only in the Occident." He simply does not know about the books and periodicals produced in China, never having been there.) And so on.

Max Weber deserves our attention not because he was a remarkable scholar and an important interpreter of the so-called "European miracle," the supposedly unique and spectacular rise of Europe long before modern times (see Chapter 5), among scholars of the early twentieth century, but because he established a distinctive tradition, even a school, in regard to the problem of explaining the uniqueness of the West, and this tradition is very active, perhaps even dominant, today. This Weberian tradition has a number of distinctive features. It is distinctive in its general approach and distinctive also in several of its concrete empirical arguments. About the general approach we have perhaps said enough already. It emphasizes ideas and values, and more broadly social factors distant from the crassly economic and technical and material. Therefore it appeals to those interpreters who tend to believe that the key factors in Europe's evolution (and non-Europe's nonevolution) were matters of political formation, family structure, religion, and, quite simply, rational thought and action.

Three of Weber's special arguments have been extremely influential among all interpreters of the "European miracle." These are three specific ways in which, according to Weber, social evolution in the West has differed from social evolution, or rather lack of social evolution, in the East (not to mention the South). The first and broadest argument is a model of the difference between a modernizing and a traditional society. Weber dwells at length on the ways in which social evolution in Europe has involved a gradual and steady distancing of society from kinship and from what he thinks of as "irrational" beliefs, such as superstition. Most of what he says in this regard is unexceptionable. But Weber contrasts this with a purely mythical model of non-Western societies. Chinese society, for instance, is supposedly held back by the fact that the clan still forms the key unit of social structure (it does not), and Chinese people are in the last analysis irrational because they have not been able to overcome a belief in magic and superstition (this is nonsense).[17] Such things go to make up Weber's model of "traditional society."

Euro-American social science of the period from about 1945 to 1965 was building what is now called "the theory of modernization," the theory that would, it was hoped, provide a formula for bringing modernity to "backward" societies of Africa, Asia, and Latin America. The key concept for this theory was the notion that "traditional societies" would be shaken out of their traditionalism and awakened to economic development by new and better ideas introduced from the West, rather as Sleeping Beauty being awakened by the Prince. This, the central model for a generation of economic development programs grounded in the theory that *ideas* would bring about change—that painful social conflict was not necessary for that change—and that the cause of poverty and backwardness was *tradition*, not oppression or landlessness, or colonialism, was in perfect harmony with Weber's theory of change and indeed made great use of that theory.

Today the theory of modernization is rather tarnished because we know that poverty is not the result of the irrationality of its victims but rather of the greed and oppressive actions of landlords, tyrants, multinational moneylenders, and the like, and also we know, as social scientists, that people of poor countries are not suffering from the Weberian malady of "traditional attitudes." They tend rather to be waiting impatiently for any opportunity that may come along to help them rise out of poverty. So the Weberian model is perhaps less popular today than it was before the Vietnam War. It is still, however, the basic model for those historians who wish to explain the "European miracle": in terms of the "rationality" of the West—innovativeness, inventiveness, progressiveness, desire to achieve, and so on—and the putative "irrationality" of other civilizations. This is the simplest of all models for justifying the superiority of Europeans. It happens to be totally false.

Weber contrasted the Occidental and the Oriental city in various ways, most of which replicated at this very large scale his ideas about the difference between traditional and non-traditional society, as well as rational and nonrational institutions. Again what we find is brilliant analysis of European urbanization, going back to antiquity, but thoroughgoing ignorance about the *real* cities of non-European areas, past and present. Weber, for instance, believed that non-European cities were, in essence, the site of political rule, whereas European cities historically were autonomous and "free," and thus were the locus of social evolution. This theory is in fact a very old one, and Weber added little to it, but he gets the credit for it nonetheless: it is "Weberian" to talk about the way in which urban society was able to lead Occidental civilization toward modernity but was unable to do so in the East, where cities remained under the thumb of empires. But if we make fair comparisons, era for era, between European and non-European cities, we tend to find, first, that many cities in both regions were quite similar down to the early modern centuries, before the industrial revolution; and second, that autonomy was greater than Weber believed in the East and much less in the West. The idea that there was "freedom" in medieval European cities is in part the error of telescoping history, projecting the characteristics of modern cities backward into the Middle Ages, and it is in part the more general diffusionist error, commented upon previously, that sees "freedom" in everything European and "despotism" in everything Oriental.

The third argument concerns landholding systems, East and West. Weber popularized, though he did not originate, the idea that the medieval manor of Europe is much closer to genuine private property, and also much more pregnant with potential for social change, than the lange estates of Asia (Africa remaining, as always, undiscussed). Weber believed that the large agricultural landholdings of the Orient were retained as property of the state or the king, while the magnates whose income came from the peasants on these estates were really not the landlords but merely temporary grantees of the estates who had been lent them in return for service to the state. Hence the estates were not real property and were contingent on continued service. In the medieval West, Weber thought, the relation between king and lord was a much more rational one, grounded in various ethical commitments, and therefore the granting of an estate to the lord was truly a grant of property, not a temporary gift of the income from that property. From this comes the terribly important Weberian contrast between European fiefs granted as (in essence) private property and Asian fiefs granted temporarily on service tenure. Said Weber: this was crucial for the development of individualism, private property, and eventually capitalism in Europe and the nondevelopment of such things elsewhere.[18] Many other thinkers, again, held to this view—Marx being one of them—but the theory is usually designated as "Weberian." However, this supposed contrast is actually not a real one. Medieval European estates were usually granted on service tenure, that is conditionally, as were non-European estates, and both tended in various regions to congeal into permanent, hence in essence private, ownership. Generalizing, the manors of Europe and non-Europe tended to have broadly common characteristics, the differences being matters that had nothing to do with greater or lesser "modernity."[19]

When all is said and done, Max Weber's theories about the supposed uniqueness of European history are wrong for two basic reasons. The special qualities that he imputed to ancient and medieval Europeans were *also* possessed by non-Europeans, who were no more traditional, no less progressive, no less rational than Europeans. And the history of society in general is much more than the history of ideas.

NOTES

1. Weber, *The Protestant Ethic and the Spirit of Capitalism* (1958a), p. 79ff.

2. Weber, *Protestant Ethic*, pp. 30–31. Also see Weber, *General Economic History* (1981); Weber, *The Religion of China* (1951), pp. 230–232.

3. Weber, *General Economic History*, p. 379.

4. Weber, *General Economic History*, p. 299.

5. Weber, *Religion of China*, pp. 231–232.

6. Weber, *Protestant Ethic*, pp. 30–31. Also see Weber, *General Economic History*, p. 379; Weber, *Religion of China*, pp. 230–232.

7. Weber, *General Economic History*, Chapter 3. On the theory of Oriental despotism, see Venturi, "The History of the Concept of Oriental Despotism in Europe" (1963); Anderson, *Lineages of the Absolute State* (1974). Also see Volume 1 of *The Colonizer's Model of the World*, Chapter 2. Oriental despotism is also discussed in Chapters 5, 6, and 8 of this volume.

8. See Romila Thapar, "Ideology and the Interpretation of Early Indian History" (1982), pp. 389–412; see also Volume 1, Chapter 2.

9. Weber, *Religion of China*, pp. 20–25; Weber, *General Economic History*, pp. 56–57. Also relevant is *The Agrarian Sociology of Ancient Civilizations* (1976), pp. 148, 157–158.

10. Weber, *The Agrarian Sociology of Ancient Civilizations*, pp. 157–158.

11. See, in Volume 1 of *The Colonizer's Model*, the section of Chapter 2 titled "Arid, Despotic Asia."

12. Weber, *The City* (1958b).

13. "[It] is fundamental to Weber's method that the individual social actor is the sole locus for the empirical production of all historical events": Ira Cohen, "Introduction" to Weber's *General Economic History*, p. xx.

14. See Chapter 8. Also see the discussion in Chapter 1 of *Colonizer's Model*, Volume 1.

15. Weber, *Protestant Ethic*, pp. 1–31.

16. Weber, *Protestant Ethic*, pp. 30–31.

17. See Kumar, "Private Property in Asia: The Case of Medieval South India" (1985), for a contrary view.

18. Weber, *Religion of China*, pp. 95–100.

19. Weber speculates here and there on other historical factors favoring Europeans. European bureaucracy is rational, while that of non-European civilizations is not: see in particular Weber's *Economy and Society* (1968). Ownership of herds was connected to the "individualism" of ancient Europeans (*Agrarian Sociology of Ancient Civilizations*, p. 37). Greek urbanization was one source of secularism and capitalism (Weber, *Agrarian Sociology*, p. 158; *Religion of China*, p. 15), and so on.

CHAPTER 3

Lynn White, Jr.:
Inventive Europeans

A book titled *Medieval Technology and Culture Change* was published in 1962 and quickly became one of the more widely read and widely discussed essays in historical explanation. The author, the late Lynn White, Jr., was an eminent North American historian.[1] The book itself is an effort to explain all of the main features of medieval social change, and to some extent modern social change, in terms of strict technological determinism. Technological determinism is the claim that new technology is the primary cause of historical change, regardless of how one tries to explain the origins of the technology itself.

The structure of White's argument is simple and straightforward. During the Middle Ages, Europeans invented, and in a few cases borrowed from others, a number of revolutionary technological traits; and, thanks largely to these technological innovations, Europe progressed first to feudalism, then advanced toward capitalism and modernity. Apart from an occasional comment in passing on the "imagination," "talent," and "dynamism" of Europeans, the argument in this book is technological determinism in the narrowest sense. Here, for starters, is a representative example. The early-medieval invention of the iron stirrup had a "catalytic . . . influence on history." It permitted a new and much more effective form of mounted warfare. This produced the phenomenon of the medieval knight. This, in turn, produced feudalism (when knights became manorial lords). And finally,

[The] Man on Horseback, as we have known him during the past millennium, was made possible by the stirrup. (p. 38)[2]

But White's crucial argument concerns productive technology, mainly in the field of agriculture. If we examine this argument in some detail, we will uncover one of the main roots of Eurocentric tunnel history in its present-day form.

A MEDIEVAL AGRICULTURAL REVOLUTION

Lynn White asserts that an agricultural revolution occurred during the Middle Ages in Europe, and in Europe alone. (He limits the venue, in fact, to northern Europe.) He believes that this revolution involved three main technological innovations: the introduction of the heavy plow; the introduction of the horse collar, and the use of horsepower, especially in plowing; and the introduction of the "three-field system" of rotation. The heavy plow, pulled by teams of (typically) eight oxen, is assigned, by White, a tentative central European origin in the sixth century, and is considered by him to have diffused throughout northern and northwestern Europe by the time of Charlemagne—and indeed "does much to account for the bursting vitality of the Carolingian realm in the eighth century" (p. 54).

White is correct in calling attention, as others before him have done, to the importance of the introduction of the heavy plow as an agricultural innovation in the wetter and colder parts of Europe. It was highly advantageous in opening the damp, heavy soils of the North European Plain, and it permitted the deep-working that was especially critical in view of northern Europe's moist weather. One can even say that it was a necessary condition for the areal spread of moderately intensive cultivation to some typical northern soils, and perhaps for the first agricultural penetration of the swampy lands that stretch eastward from Holland across much of northern Germany and beyond, and the difficult hillside soils of rainy regions such as some western and northern parts of the British Isles. Therefore, the heavy plow had much to do with the overall increase in medieval Europe's agricultural production, because it extended the cultivated area and increased areal productivity.

Notice, though, that we can accept these generalizations without falling into the clutches of technological determinism. The heavy plow was not a radical departure from the principles embodied in older, lighter types of plows, long used in southern Europe (where soils are generally drier and lighter, and deep-working is not needed and usually is poor practice). Hence *any* causal process that would lead the northern peasants to try to expand the acreage under cultivation, and to increase production on wetter soils, would certainly lead them to make the relatively minor technological adjustments that would adapt known plow technology to northern soils. Lynn White, however, begins his argument with the tool, not with the culture. He quite fails to notice the social forces of feudalism that were leading the landlord class to demand ever larger surpluses from peasants and thus force the peasants to increase their production in any way possible, as well as the social forces that were impelling landlords in many regions to try to enlarge their estates by encouraging pioneer settlement (for their own, not the peasants', benefit), much of it in the wet margin of agricultural settlement across northern Europe; and, finally, the quite reasonable desire of the peasants themselves to grow more food for themselves, all other forces aside.

For White, the plow was the cause and social change the effect. And the effect, in his view, was altogether revolutionary. Thanks to the adoption of the heavy plow, says White, there was, to begin with, a tremendous growth of population. Then there was a changeover to the "open field" system of cultivation (the system in which large, undivided fields were plowed in strips, each family then planting one or more strips within the field); and this technological innovation, in White's view, led to new "communal" patterns of human cooperation—note the explicit denial that there were preexisting communal patterns in agriculture, as postulated by Marxists and by most anthropologists—and thence to the manorial system as a whole. According to White, the "essence of the manorial economy" was joint exploitation of the large open fields, hence cooperative decision making by villagers—neglecting the fundamental fact that the manor belonged to the lord, not the villager (p. 44). Thanks to the heavy plow, the social structure was very different. More important still, there came then "a change [in] the northern peasants' attitude towards nature, *and thus our own*" (p. 56, my emphasis). This last entails a curious argument:

> From time immemorial land was held by peasants in allotments at least theoretically sufficient to support a family. . . . Then in northern Europe, and there alone, the heavy plough changed the basis of allotment: peasants now held strips of land at least theoretically in proportion to their contribution to the plough-team. Thus the standard of land distribution ceased to be the needs of the family. . . . No more fundamental change in the idea of man's relation to the soil can be imagined: once man had been a part of nature; now he became her exploiter. (p. 56)

Here there is a huge leap from the early-medieval plow to the modern ("our own") attitude toward nature, but, this aside, the evidence supports none of White's reasoning. There are two main objections, each alone sufficient to bring down the entire thesis. We do not have enough historical and archeological evidence yet to examine the matter on a worldwide scale, but the evidence we do have suggests that (1) the heavy plow was not invented in Europe (or only in Europe) and (2) it arrived in northern Europe long before the time postulated—of necessity—by White: the time when population was beginning to grow after the Dark Ages and the feudal manor was beginning to establish itself widely.

In northern India, heavy plows drawn by as many as twenty-four oxen were in use by the fifth or sixth century before Christ. Heavy plows, evidently with moldboards and coulters, were present in Roman Britain, and no doubt elsewhere in northern Europe at the same time.[3] Likewise, the open-field system, which, according to the Orwins in their classic work on the subject, *The Open Fields*, was a feature of the methods of cultivation that stretched across the plains of Europe and Central Asia in primitive times.[4] Thus, Lynn White's argument is countered by comparative historical evidence. In particular, there is nothing peculiarly "European" about European ideas toward nature—if they have anything to do with the adoption of the heavy plow. But they do not.

This brings us to the second objection. The causal chain does not bear weight. There cannot have been much connection between the adoption of deep plowing and any significant change in social organization (let alone attitudes toward nature), other than those changes that obviously accompanied the general expansion of settlement and population in that region and era. Northern European villages displayed no greater degree of cooperation than southern villages, which used lighter plows, with smaller teams, because these were more suited to the lighter, drier southern soils. And the heavy plow, in any case, would not lead to the "communal" effects attributed to it by White, who requires us to accept at least three false premises: that prior farming was essentially a matter of individual—even private—holdings and work; that many families had to band together to support one plow-team (the Domesday Book gives plow-team-to-household ratios of between 2:3 and 3:5 for England—thus not much more than one household per team)[5]; and that the manorial system, the central feature of feudalism, was mainly a result of technological innovation.

It seems much more reasonable to view the heavy plow as one element in a complex of changes that, before genuine feudalism appeared in northern Europe, accompanied the successful effort by northern peasants to expand their frontier of cultivation. Undoubtedly the plow spread more widely during the feudal period, but this presumably reflected the joint causes of population growth (associated with general pacification) and intensified feudal exploitation—that is, increasing need for food and increasing demand by landlords for surplus—not the mystical force of technological progress. White's argument, as one European scholar puts it, betrays an "enthusiasm for 'progress' to the point of forgetting the need for historical verification."[6]

The second revolutionary advance is labeled by White "the discovery of horse-power" (p. 57). Horses had been around, of course, for some time. The essential innovation, according to White, was the modern horse collar, which he believed was probably "the product of a slow development in the Occident," although he concedes that there is some evidence that it may just possibly have diffused from central Asia. (Here I must pause to make a general observation about the mode of explanation employed not only by Lynn White but also by most Eurocentric historians who argue from the position of technological determinism. Where possible, they find Europeans inventing the technique. Where necessary, they concede an extra-European source and then move, smoothly and swiftly, to the argument that Europeans alone figured out how to put the thing to good use. The type example of this sort of reasoning is gunpowder. The Chinese invented it, as every schoolchild knows, just to make firecrackers with. The Europeans put it to use in guns. More on this in later chapters.) In any event, the modern horse collar was fully evolved and widely used in northern Europe by the ninth century. The argument then proceeds from this putatively European innovation to the conclusion that it transformed the agriculture and the grain transport of northern Europe, roughly in the twelfth century, by permitting horses to replace oxen in pulling heavy plows and wagons. White points out, correctly, that a horse pulls about the same weight as an ox, but does so about fifty percent faster. From this fact he draws the most awesome conclusions. There was, first of all, a major increase in agricultural production. The intensity of commerce rose, because transport by horsepower was (according to White) vastly cheaper than by ox-power. Villages became much larger, almost townlike, because now there could be a larger radius of travel from home to field and therefore much more cultivable acreage lay within reach of the village. The presumed enlargement of villages yielded the "virtue of a more 'urban' life," permitting the village to have a church, a tavern, and a school (now boys "could learn their letters," p. 67), as well as greater commerce with the outer world and intensified communication—"news from distant parts"—a transformation, overall, of profound importance:

Deep in the Middle Ages this "urbanization" of the agricultural workers laid the foundation for the change in Occidental culture from country to city which has been so conspicuous in recent centuries. It gave the peasantry of northern Europe psychological preparation for that great shift and perhaps enabled them to build up attitudes and spiritual antibodies which reduced the social shock of subsequent developments. (p. 68)

All this from one innovation: the horse collar.

But the modern horse collar was in use some centuries earlier in China. Needham suspects that it was widespread in Eurasia from an early date, and may have been invented for harnessing camels.[7] Thus, to tie this one tool to a "great shift" in "Occidental culture" is just wrong—wrong to make the causal connection, wrong to neglect Oriental culture.

And again there is a second objection. To begin with, the presumption that the horse held an advantage over the ox in plowing, and even in transport, is widely disputed.[8] The horse was more efficient, but also much more costly in upkeep: among other things, it called for the assignment of cropland to oat and hay cultivation for feed, meaning more farm labor input and less land for food crops. In England the horse apparently did not replace the ox during the Middle Ages, but even if perhaps it did in some other areas the causal argument from horse (or rather horse collar) to such things as the commercialization and urbanization of life (and literacy, and so on) is unacceptable. In many parts of the world outside of northern Europe, villages were much larger in that period, and also productivity per acre was much higher, yet oxen and water buffalo were used instead of horses. In China, it may be noted, large-scale grain transport at that time was mainly via canal boats—far cheaper than using horses—yet canals were as cheap to build, and easier to maintain, in the North European Plain than in China but, even so, were neglected in the former area until the Middle Ages had passed. And so on.

Finally, Lynn White calls our attention to the three-field rotation, which "bursts upon us in the late eighth century after its invention in northern France, then spreads rather slowly, reaching England in the twelfth century" (p. 69). Part of this argument is now rather conventional: even in high school history we learned that the three-field system was a great advance over the older two-field system. We learned that it increased the crop-to-fallow ratio from (roughly) 1:1 to 2:1 and spread agricultural work over a slightly longer period in the year. White is not content, however, to argue an increase in productivity per acre (and, he thinks, per worker). He adds that oat production would now be more widespread, because more cropland presumably equals more fodder crops, and thus there would be still greater use of horses. "But people likewise were shaped by the new resources"—the leguminous crops, like peas and beans, which were normal components of the three-field rotation. (This section of White's book carries the heading "The Three-field Rotation and Improved Nutrition.") At this point, White's technological determinism reaches its zenith:

> It was not merely the new quantity of food produced by improved agricultural methods, but the new type of food supply which goes far toward explaining, for northern Europe at least, the startling expansion of population, the growth and multiplication of cities, the rise in industrial production, the outreach of commerce, and the new exuberance of spirits which enlivened that age. (p. 76)

In short, says Lynn White, Jr., the Middle Ages "were full of beans" (p. 76).

None of this, least of all the pun, can be taken seriously. To begin with, White's argument (Malthusian by his own admission) that population had earlier been held down by a kind of starvation ("a diet overloaded with carbohydrates," p. 75) is grounded neither in historical nor in scientific evidence. Farming populations using the two-field rotation were *not* protein-poor, since (1) legume cultivation long antedated the three-field system in Europe, (2) grains contain protein, and (3) fruits, animal products, and nonleguminous vegetables had been widely consumed in earlier periods.

More importantly, most European scholars, and particularly historical geographers (who tend to know something about agriculture), are generally agreed that the three-field system was very different from White's image of the system. It was doubtless an important modification of technique in part of northern Europe. It did not diffuse to some other regions mainly because in some places it was not always an advance over other rotations, including, in some areas, the two-field system. In some parts of Europe it is known to have been adopted and then abandoned later in favor of the two-field system. The picture that is painted for us by these scholars shows most northern-European farmers adopting a new and, for them, good innovation that moderately improved their productivity, while others (equally well-nourished) found this innovation unsuitable and rejected it. This is a far cry from White's cornucopia, out of which flow an "exuberance of spirits" and the rest. The three-field rotation was not the most intensive and productive system in Europe in that period: essentially continuous cropping seems to have been practiced in parts of northern Italy, the Netherlands, some areas of "infield-outfield" agriculture (some of the manured infields), and it seems reasonable to surmise that, in many European regions where soil nutrient status and soil water relations were good, continuous cropping could be carried on indefinitely without soil loss.[9] The three-field rotation was not preferred in some areas where fallow was needed for grazing (since it reduced the fallow and increased the cropland). It did not suit certain poor soils that needed longer fallows. The picture, in a word, is complex. Furthermore, continuous fallow-free rotations with and without irrigation, some of them highly intensive and quite intricate, were found in other parts of the world, so any advantage claimed for the European three-field system has to be conceded to non-European farmers as well.

But the crux of the matter is, again, causality. Adoption of the three-field system was an example of the ordinary sort of intensification process that occurs throughout the world in agricultural systems where peasant farmers do not have enough land and where landlords exact an ever increasing demand for surplus. Farmers everywhere are inventive.

TECHNOLOGICAL INVENTIVENESS

Among all of the narrow-minded ways of looking at history today, technological determinism is the one most congenial to Eurocentric tunnel vision. It has the appearance, the illusion, of cold-blooded scientific fact: "X was invented *here*, on *this* date, by *these* people, and produced *these* effects." Thus, in talking about matters of technology one can always deny that ethnocentrism enters the picture. One is talking about *facts*. Can you argue with *facts*? Compare this kind of reasoning with those other forms of determinism that rest in concepts hard to define and hard to locate: arguments rooted in value systems, social organization, and the like. If one tries to compare cultures on these latter attributes, the charge can always be made that the criteria and scales are ethnocentric. It is hard, to say the least, to prove that one value system, one family system, one religion, or the like is more highly developed, more modern, than another. It is not hard at all to prove that one tool produces more food, or more cloth, or more enemy casualties than another. Hence the peculiarly convincing quality of arguments grounded in technology; and hence the preference for this kind of argument, and evidence, in efforts to prove the superiority of Europe at all times, and most crucially in the Middle Ages.

But technological determinism is a paper tiger. Material technology is not imbued with life; it does not evolve like plants and animals, obeying inner laws. A new tool arises not by mutation from an ancestral tool: its appearance has to be *explained*. And to explain it one must shift the entire basis of the argument to other grounds. Tools are material things; they lie at the interface of culture and nature. Therefore, one can explain their origin in terms of cultural forces, or cultural forces interacting with environmental qualities. But one cannot do so in terms of other material tools.

Thus, technological determinism must *always* dissolve into something else, some argument in which technology is not the explanation, but the thing-to-be-explained. And the explanation, in turn, may prove to be some argument about certain "values," certain kinds of social structure, certain religions, and so on, that are claimed to be conducive, or not conducive, to technological invention and innovation. This is the key point, and its significance for explanations about Europe's medieval cultural change will become apparent in a moment.[10]

Lynn White holds to a very definite theory as to why medieval Europeans invented the technological traits that, as he thinks, propelled them, uniquely, forward to modernity. This theory is still the basic, root belief that Europeans are naturally more inventive and innovative and creative than non-Europeans, and White believes he knows *why*.

TECHNOLOGY, ECOLOGY, RATIONALITY, RELIGION

A few years after the appearance of *Medieval Technology and Social Change*, Lynn White published a very important and very influential article titled "The Historical Roots of Our Ecological Crisis."[11] The article appeared in *Science* just during the awakening of the ecology movement in the United States and the awareness that there, indeed, *was* an ecological crisis. Said White: to cope with the crisis, we must begin by "looking, in some historical depth, at the presuppositions that underlie modern technology and science."[12] What do we find? "One thing is so certain that it seems stupid to verbalize it: both modern technology and modern science are distinctively *Occidental*."[13] "From the eleventh century onward, the scientific sector of Occidental culture has increased in a steady crescendo,"[14] and "by the late 13th century Europe had seized global scientific leadership."[15] Therefore:

> Since both our technological and our scientific movements got their start, acquired their character, and achieved world dominance in the Middle

Ages, it would seem that we cannot understand . . . their present impact upon ecology without examining fundamental medieval assumptions and developments.[16]

The "developments" are then briefly illustrated with a summary of White's theory of the heavy plow (which we have already discussed), and White next turns to the "assumptions." These are in fact strongly reminiscent of Max Weber, although White does not cite Weber (doubtless because both he and Weber are transmitting a widely held theory). The "assumptions" are matters of rationality and religion.

> [Technological] novelties seem to be in harmony with larger intellectual patterns. What people do about their ecology depends on what they think about themselves in relation to things around them. Human ecology is deeply conditioned by beliefs about our nature and destiny—that is, by religion.[17]

But the category "religion" is quickly reduced to the category "Western Christianity." There are, in fact, two assumptions. One is the "Judeo-Christian teleology," which, says White, is the source of the Occident's "faith in perpetual progress," a faith that, he says, is not found in other religions and is transmitted only through the Western forms of Christianity.[18] The second is "an Occidental, voluntarist realization of the Christian dogma of man's transcendence of, and rightful mastery over, nature"—in essence, the separation of Man and Nature. To a Christian (or rather Western Christian), nature is inert, valueless. It is idolatry to "assume spirit in nature." "To a Christian a tree can be no more than a physical fact." "Our science and technology have grown out of Christian attitudes toward man's relation to nature." Therefore, "Christianity bears a huge burden of guilt" for our ecological crisis.[19] And White concludes that, "since the roots of our trouble are so largely religious, the remedy must also be essentially religious."[20]

So White seems to be a religious determinist, not a technological determinist. Or rather both: one behind the other. In his view, one form of the Christian religion produced the uniquely scientific and technological inventiveness of the Europeans, and the outpouring of new technology during the Middle Ages then led to Europe's modernization.

Several very serious errors can be discerned in this view. One of the most obvious is White's lack of attention to the teachings of religions other than Christianity. He resurrects an old myth of Eurocentrism to the effect that ancient pagans and followers of modern religions other than

Christianity (and Judaism) are unable to fully separate Man from Nature: they share the primitive view that spirits reside in all things.[21] Students of comparative religion are aware, however, that Asian and African religious thought has strains somewhat parallel to the Christian, including even, here and there, dualism and a sort of materialism.[22]

Secondly, White commits the sin (for a historian) of telescoping history: in this case taking modern facts and traits and attitudes and imputing them falsely to times past. The attitude of modern Europeans toward nature is essentially a product of the period since the rise of capitalism and more importantly since the industrial revolution. The rise of capitalism generated a sort of "commoditization" of the natural world, which came to be imbued with, so to speak, price in place of value. Attitudes toward nature thus shifted, and did so rather suddenly. White to the contrary notwithstanding, medieval European thought tended to place humanity clearly *within* nature. Medieval Catholicism taught the doctrines of plenitude, of unity, nonseparation, the idea of the "great chain of being" according to which God left no holes in nature, no gaps in the chain of being, and the idea that entities other than human beings had souls of a sort was widely entertained and presumably not considered blasphemous.[23] We see, here, that White's category "Western Christianity" is really "post-Reformation Christianity." This is erroneous in its view of medieval Christianity, and thus it is erroneous in its sweeping references to "the Judeo-Christian teleology" as being the roots of our present-day ideas about the separation of Man and Nature, about science, and the rest.

Modern science and technology really emerge from the period of early capitalism.[24] This brings us to the third serious error. Lynn White, like Max Weber and countless other Eurocentric thinkers, past and present, believes that the triumphs of modern European culture, including science and technology, do not reflect recent historical shifts, like the rise of capitalism and the expansion of Europe after 1492, but are rooted deep in the European psyche, culture, and history.[25] In the Middle Ages, as later, Europeans were superior, just in the ways they are now. They have been superior always or since ancient times, whatever the reason (race, religion, environment, culture). In fact, that belief is mistaken, as we show in this book, and where it draws on empirical history, it is a Eurocentric tunnel history that quite ignores the past of non-European civilizations. Enough is now known about the history of Chinese, Indian, and Islamic science to make it very clear that European science and technology was in no way superior to them prior to the modern period—after the Middle Ages had ended, capitalism had begun to rise, etc. Each civiliza-

tion had certain special flairs in one or another branch of science technology, but none was truly in advance of the others. Indeed, most of the
ideas were *shared*. Massive crisscross diffusion of ideas, traits, inventions,
skills, and skill-bearers took place throughout the Eastern Hemisphere
during the Middle Ages, and all of the major civilizations tended, in this
aspect of culture, to share a common evolution.

White's notion that modern European ideas about perpetual progress
are rooted in ancient doctrine, in a "Judeo-Christian teleology," are erroneous for roughly the same reasons. These are modern ideas, as most
scholars agree. Medieval Europeans tended to see God as perfect and
therefore the Creator of a world that could not become better. Admittedly, this denial of the possibilities of progress was not universal, but it is
quite incorrect to imbue medieval Europeans with the modern belief in
"perpetual progress." This belief is a product of progress itself—of the
growth of capitalism, the improving standard of living of European people, and other attributes of modern times. Moreover, the idea that non-
Europeans do not believe in progress, either as a matter of culture or as
one of religion, is quite false. It is the idea, abundantly criticized throughout the present volume, that non-Europeans are "traditional," "stagnant,"
not motivated toward change, and the rest.

Just in passing I would like to dispute Lynn White's notion that the
belief in a dualism between Man and Nature, spirit and matter, mind and
body, and so on, is peculiarly and conspicuously European and Christian.
It is basically the Cartesian and post-Cartesian strain in modern European thought, and in spite of much attention to *some* among the philoso-
phers of antiquity (most conspicuously Plato), dualism was *not* characteristic of European thought in prior epochs.

NOTES

1. Lynn White, Jr., *Medieval Technology and Social Change* (1962). Numbers given
in parentheses in the text of this chapter refer to this work.

2. White, *Medieval Technology* p. 38. Whether or not the iron stirrup was invented
in Europe, hardened stirrups using wood were widespread in Eurasia for many centuries
before the iron stirrup arrived in Europe. There is not much difference between wood and
iron in this context, so White's theory about "the Man on Horseback" and the rest is
completely wrong. See Needham and collaborators, *Science and Civilization in China: Vol.
4, Part 2* (1965).

3. Kosambi, *Ancient India: A History of Its Culture and Civilization* (1969), p. 89;
Sharma, *Light on Early Indian Society and Economy* (1966), p. 57.

4. Orwin and Orwin, *The Open Fields* (1967), Chapter 3.

5. Smith, *An Historical Geography of Western Europe Before 1800*(1969), p. 203. H. C. Darby, in his definitive work *The Domesday Geography of Eastern England* (1952), p. 379, estimates a rural population of 88,000 as against 21,000 plow-teams in the Eastern Counties, which suggests that there may have been roughly one team per household.

6. Titow, *English Rural Society: 1200–1350* (1969), p. 37.

7. Needham, *Science and Civilization in China: Vol. 4, Part 2*, pp. 304–328.

8. Parain, "The Evolution of Agricultural Technique," pp. 143–145. Also Titow, *English Rural Society*, op. cit., pp. 37–39.

9. The older idea that only shifting agriculture, not permanent cultivation, was possible in the absence of fertilizer application is now known to be false, for tropical areas as well as temperate ones. Volcanic soils, well-drained alluvial soils, some limestone soils, and many other soils typically support permanent cultivation. See Blaut, "The Nature and Effects of Shifting Cultivation" (1962) and "The Ecology of Tropical Farming Systems" (1963). I suspect that many regions south of the Alps, in the Danube basin and elsewhere had some permanent cropping during the Middle Ages. For discussions of the three-course rotation and other systems contemporary with it, see Smith, *An Historical Geography of Western Europe*, pp. 203–218; Pounds, *An Historical Geography of Europe: 450 B.C.–A.D. 1330*(1990), pp. 366–379; Parain, "The Evolution of Agricultural Technique," pp. 136–142; Slicher Van Bath, *The Agrarian History of Western Europe: A.D. 500–1850* (1963), pp. 58–62.

10. And there are other problems; other reasons for entertaining doubts about assertions (like Lynn White's) concerning technological change and its significance for cultural change in the Middle Ages and in general. Tools are not always "hard facts" for the historian: often they go unmentioned in documentary sources because they are mundane and their inventors and users usually are common people. And there is nothing self-evident about the significance of a new tool for culture change. It may have been invented, literally, as a cultural tool.

11. Originally published in *Science*, March 10, 1967; reprinted in Lynn White's *Machina Ex Deo: Essays in the Dynamism of Western Culture* (1982).

12. White, "The Historical Roots of Our Ecological Crisis" (1982), p. 79.

13. White, "Historical Roots," pp. 79–80.

14. White, "Historical Roots," p. 82.

15. White, "Historical Roots," p. 82.

16. White, "Historical Roots," pp. 82–83.

17. White, "Historical Roots," p. 84.

18. White, "Historical Roots," p. 85.

19. White, "Historical Roots," p. 90.

20. White, "Historical Roots," p. 93.

21. In its typical form, this theory ascribes such views to a kind of childish stage in cultural evolution: children, ancients, modern primitives, modern non-Judeo-Christians, and mentally defective adults all share an inability to fully distinguish the human being from the environment, and impute spiritual qualities to the latter. Sometimes women are added to the list. See my essay "Diffusionism: A Uniformitarian Critique" (1987b) and *Colonizer's Model*, Volume 1.

22. See, for instance, Chattopadhyaya, *Lokayata: A Study in Ancient Indian Materialism* (1967).

23. See, for example, Lovejoy, *The Great Chain of Being* (1936).

24. White: "In the present-day vernacular understanding, modern science is supposed to have begun in 1543, when both Copernicus and Vesalius published their great works. [But the] distinctive Western tradition of science, in fact, began in the late 11th century" (*Medieval Technology*, p. 82). The "vernacular understanding" is more-or-less correct: modern science perhaps begins with the era of Copernicus and Galileo—that is, after 1492. Prior to that time European science was progressing no more rapidly than science in other continents, and was not "distinctive."

25. It is a common misconception that Max Weber sees modern science and technology as beginning with the Reformation. See Chapter 2.

CHAPTER 4

Robert Brenner:
The Tunnel of Time

EURO-MARXISM

Robert Brenner is a Marxist historian, a follower of one tradition in Marxist scholarship that is as Eurocentric as most conservative positions. I cannot here offer an explanation for this curious phenomenon: a tradition within one of the most egalitarian of all sociopolitical doctrines, yet a tradition that, nonetheless, believes in the historical superiority (or priority) of one community of humans, Europeans, over another, non-Europeans. Eurocentric Marxists are not racist, nor even prejudiced, although most of them believe that Europeans have always been the leaders in the forward march of history and that Europe is the fountainhead of civilization, the main source of innovative social change. For these scholars, the origins of capitalism are European. Capitalism's further development consisted of an internally generated process of improvement within its classic homeland, the European world. Colonialism was not significant for capitalism, was rather a marginal process, a temporary aberration or diversion or sideshow, not a vital need of the system as a whole, which evolves in response to internal laws of motion.

This point of view is basic Eurocentric diffusionism. It is also tunnel history: a form of tunnel vision that (as we saw in Chapter 1) tries to explain the rise of capitalism and the rise of Europe by looking only at prior European facts—looking, as it were, down the European tunnel of time, ignoring the history of the world outside of Europe both as cause of change within Europe and as the site of historically efficacious change in its own right.[1] The Euro-Marxists—as I will call the scholars of this tradition—accept this view, and so they are Eurocentric diffusionists. To this extent, they agree with their mainstream colleagues about the rise of Europe, of capitalism, of modernization, of industrialization, of democracy: basically all of it is European.

Euro-Marxism went into eclipse during the period when liberation movements were decolonizing most of the world. In this period, the idea that the colonial or Third World has been, and is, unimportant in social development was not popular among Marxist scholars. After the end of the Vietnam War, however, this point of view became again popular, and indeed became the Marxism most widely professed in European and American universities. Today we witness the curious phenomenon that Euro-Marxists are quoted with approval by conservative scholars who can use them to show that "real" Marxist scholarship supports some of the same doctrines, theoretical and practical, that conservatives do.

Robert Brenner is one of the most widely known of Euro-Marxist historians. His influence stems from the fact that he supplied a crucial piece of doctrine at a crucial time. Just after the end of the Vietnam War, radical thought was strongly oriented toward the Third World and its struggles, strongly influenced by Third World theorists such as Amilcar Cabral, Franz Fanon, Che Guevara, C. L. R. James, and Kwame Nkrumah, and thus very much attracted to theories of social development that tended to displace Europe from the pivotal position claimed for it by Europeans, the center of social causation and social progress, past and present. Euro-Marxism of course disputed this, and Euro-Marxists, while strong in their support of present-day liberation struggles, nonetheless insisted as they always had done that the struggles and changes taking place in the center of the system, the European world, are the true determinants of world historical changes; socialism will rise in the heartlands of advanced European capitalism, or perhaps everywhere all at once; but socialism will certainly not arrive first in the backward, laggard, late-maturing Third World.

What was badly needed at this juncture was a strong Euro-Marxist theory of the *original* rise of capitalism, a theory demonstrating that capitalism and modernization originated in Europe and evolved thereafter mainly in Europe, with little influence from the non-European world and colonialism. The crucial questions were matters of medieval and early-modern history, to prove that Europe was the source of innovation back in those times, and so the modern European world is still, by implication, the main source of innovation. Robert Brenner supplied such a theory in two long essays in 1976 and 1977, followed by another in 1982.[2] These essays are among the most influential writings in contemporary Marxist historiography, influential among conservatives and Marxists alike.

The first of Brenner's long essays, "Agrarian Class Structure and Economic Development in Pre-industrial Europe," appeared in the history journal *Past and Present* in 1976. It was presented as a Marxist critique of conventional, conservative theories concerning the origins of capitalism in Europe (the rest of the world ignored), particularly those theories that focused on demography and on trade and urbanization as prime causes. The paper provoked a number of replies in the same journal, and Brenner issued a comment-in-reply in this journal in 1982 ("The Agrarian Roots of European Capitalism"). The whole exchange was then published as a volume, *The Brenner Debate*, in 1985.[3] Other comments on Brenner's essays have appeared in various journals from time to time, and are still appearing.[4]

In 1977 Brenner published a very different sort of essay in *New Left Review*. In this paper, "The Origins of Capitalist Development: A Critique of Neo-Smithian Marxism," he restated his theory about the European origins of capitalism and then leaped forward into the twentieth century to use this theory as a weapon against what he called "Third-Worldist" deviations in modern radical scholarship. The main targets of this attack were three well-known scholars, Andre Gunder Frank, Paul Sweezy, and Immanuel Wallerstein. These three were at the time among the most widely read exponents of a theoretical perspective that emphasizes the crucial significance of colonialism and neocolonialism, and the struggle against it, in modern history and today; Frank's view, a form of "dependency theory," Wallerstein's view, called "world-systems theory," and Sweezy's rather more traditional antiimperialist Marxism, all differed from one another in some respects but held in common the proposition that events outside of Europe have been crucial in social development since before the rise of capitalism, and the Third World is thus crucial in the struggle for socialism.[5]

To answer this argument, Brenner said, in essence: the world outside of Europe has not been important for social development. It played no role in the original rise of capitalism. It did not become underdeveloped as a result of European imperialism. And too much enthusiasm for Third World struggles today will favor meaningless reformism in the Third World and will hinder, not help, the struggle for socialism in the main arena: the West. Brenner now labeled his opponents as followers of Adam Smith rather than Marx, in their thinking about the forces of historical change, past and present. Frank, Sweezy, Wallerstein, and those who agreed with them were "Neo-Smithians."

The *New Left Review* paper, unlike the essays in *Past and Present*, was a polemic—and an effective one. Euro-Marxists and Eurocentric conservatives give much credit to this essay for what they view as the demise of dependency theory and the decline of Third World-oriented approaches in the European academy. According to M. Cooper, Brenner showed that "Sweezy et al. have put forth nothing but a restatement of Adam Smith's mechanistic and deterministic analysis of the transition from feudalism to capitalism."[6] According to John Browett, "the age of the radical-liberal dependency formulations has come to an end," thanks in considerable degree to Brenner's critique.[7] According to Alan Macfarlane, Frank, Sweezy, and Wallerstein have been "demolished" by Brenner.[8] And Brenner's critique also came to the rescue of standard economic development theory at a time when its diffusion-of-economic-modernization formula was being cast aside in favor of anti-colonial, anti-foreign, and socialist strategies, most of them justified by Marxism. The development theorists cited Brenner and announced: Marxism is on *our* side.[9]

Brenner's theory is important in several ways that are of central concern for us in the present book. It is one of the most influential and widely cited theories in this genre, in mainstream as well as Marxist historiography. (It is used and cited, for instance, by Eric Jones, John Hall, and Michael Mann, whose views we will discuss in later chapters.) It can stand as a token for Euro-Marxist historiography, an important school within the larger universe of Eurocentric historiography, and in particular it can illustrate, perhaps better than any other text, how thoroughly Eurocentric some Marxist scholarship can be, in spite of its claim to universality. And, finally, Brenner's is a fairly coherent theory about the medieval rise of Europe and capitalism and is presented by him in some detail, so it provides us with an opportunity to thoroughly dissect one such theory.

BRENNER'S THEORY

Brenner's theory is an attempt to explain why capitalism arose and why economic modernization began—all in medieval Europe. Nowhere in the three long essays does he so much as mention medieval Africa and Asia.[10] So, before we consider the theory itself we should observe its geographical perspective. Brenner is a pure tunnel-historian: the only facts worthy of mention in explaining the medieval rise of Europe and capitalism and the further progress of capitalism and economic development in the sixteenth century are European facts. We will return to this point later.

Brenner wants to show that the rise of capitalism came about as a result of social transformations in rural Europe in the late Middle Ages. The transformations he speaks of are changes in the class structure, and the dynamic process is class struggle. He begins with a description (not very controversial) of the social nature of feudal Europe during the High Middle Ages, the period of centuries (roughly the eleventh through the thirteenth) during which the feudal system in France and England reached a kind of zenith of development, while signs of relative prosperity (or lack of crisis) were evident in the countryside and towns and commerce were expanding strongly. During this period the basic class structure consisted mainly of unfree peasants, typically serfs, and a ruling class of feudal lords who owned the land and held truly life-and-death control over the peasants.

The serfs, and some of the other unfree or semifree peasants, typically performed mainly three classes of service from which the ruling class obtained its income. They were required to give unpaid labor service on the lords' demesne, a large and organizationally unified farm. They were forced to give to the landlords a portion of their own production, either in the form of the agricultural products themselves or in the form of cash rent and fees, on small farms that they worked with family labor, the land being part of the landlord's estate. And they were forced to supply labor for other things such as some artisan work, domestic service, military service, and so on. Peasant farmers did not relinquish their labor time, their production, and indeed their lives without struggle, so there was constant struggle between the producing class and the ruling class, a struggle that took both mild forms, like holding back deliveries of surplus, and radical forms, like peasant revolts. Brenner wants to demonstrate that feudalism collapsed and transformed itself into capitalism as an eventual result of this kind of class struggle that pitted feudal landlords against peasants. He wants to demonstrate at the same time that other explanations for the transformation of feudalism into capitalism are not correct. Principally, he attacks two other theories, the two that are most popular among economic historians, both conservative and Marxist, who tend to claim that "economic" factors were the main cause of the decline of feudalism and the rise of capitalism.

One of the two theories is an argument from demography to economics to society, and it rests in the Malthusian theory of population change, the theory that: (1) people cannot control their own reproductive behavior, and so they have as many children as they can; (2) this leads to overpopulation, that is, hunger and want; (3) an adjustment then takes place, with a good share of the population killed off, until again there is a rough balance between the number of people and the resources to feed them; at which point (4) the people again have too many babies and the cycle begins anew. Brenner shows how (some) conventional economic historians use neo-Malthusian arguments to explain the decline of feudalism in roughly the following way. They claim that the period of prosperity during the High Middle Ages led to rapid population growth, and this continued until peasant populations had outstripped the resources available to feed them adequately: the available productive land, the known technology, and so on. This produced a "Malthusian demographic crisis" early in the fourteenth century, which was then exacerbated by the Black Death in about the middle of that century. So, after about 1350, Europe's population declined drastically. This then produced a situation in which the landlords had too few peasants to provide them with adequate income, so they were forced to free the serfs and generally offer better living conditions to the peasants, who thereafter became free tenant farmers, paying their rent in cash. In a typical version of this theory, the end of serfdom in much of western Europe during the period 1350–1400 was really the beginning of the rise of capitalism, because the tenant farmers now, on the one hand, were participants in a commercial economy, selling their produce to obtain the money to pay rent in cash, and on the other hand had larger farms (because the rural population was drastically lower) and so had greater resources for the accumulation of capital. Always, other factors were considered to have been at work, urbanization being one of them. But the "Malthusian crisis" of the fourteenth century is considered to have been a basic cause of social change.

Brenner's attack on this theory is surprisingly timid, given his agenda. He says only the following. Demographic factors *are* important. The Malthusian model is not *entirely* wrong.[11] But demographic crisis does not explain the crisis of feudalism in Britain and France around 1350–1400. Why? Because the lords could have somehow continued to extract their income from the peasants without freeing the serfs. Brenner insists that the essential reason the serfs were freed was the success of their own class struggle—revolt and flight—during that period. Depopulation remains, nonetheless, as a factor. Brenner then adds, correctly, that feudalism as a system did not die when serfdom ended, so the demographic crisis therefore does not explain its demise.

Brenner's main attack is directed at another theory, one that explains the transition to capitalism in terms of increased commercialization and trade. Brenner's focus of attention on this theory very clearly ties in with his parallel agenda for the early-modern period (and, indeed, for the present day): to show that the non-European world had no important role in the post-1492 rise of capitalism by arguing that its role was limited to commerce and trade—not production and class struggle—and commerce and trade are not basic to capitalism either during the Middle Ages or thereafter. (I return to this point later.)

The trade-and-commercialization theory claims, broadly, that feudalism declined and capitalism rose because a change took place in the economy of western Europe from perhaps the tenth century onward, a change leading toward greater commercialization of the rural economy and greater trade, both local and long-distance. Feudalism in the early Middle Ages had displayed, in this view, a relative absence of commerce and trade, indeed a relatively subsistence-oriented economy. The feudal estates are seen in this theory as having been basically nonmonetized, with the serfs providing, with their labor, nearly all of the lord's goods and services. However, as feudalism evolved, there came a gradual change, both on and off the lords' estates. (This change did indeed occur; the problematic question is how unimportant trade and commerce were early in this period.) Towns and cities grew, commodities moved along new trade routes and were exchanged in new markets, production on the lords' demesne farms reoriented itself somewhat toward sale of the products raised, and peasants themselves began, more and more, to sell the commodities raised on their holdings, eventually reaching the level of commercialization that permitted peasant rents to be paid largely in cash. The increase in commercialization of agriculture and in trade was also accompanied by a growth in the urban economy. Overall, then, the change involved a gradual evolution toward a market economy.

The change occurred mainly for two reasons, in this theory. One reason was the reestablishment of trade connections between northern Europe and Mediterranean Europe, and between the latter and the Near East. The second, related, reason was the growth of towns. As feudalism evolved during a relatively peaceful epoch, as compared to the antecedent one, new towns appeared in the countryside, old towns became reinvigorated, and there began a slow and irregular rise in urbanization throughout western Europe; which was also accompanied by the growth of trade, partly because the towns became nodes in regional trading networks and partly because they produced certain kinds of products that were exchanged for agricultural products. The growth of the market economy, in this theory, led very smoothly and naturally to the decay of feudalism and the rise of capitalism.

Underlying this theory is a model of the human actor as somehow a natural capitalist: when confronted with a chance to "truck and barter," people will do so. The increased economic opportunities led very naturally to a sloughing off of the noneconomic constraints associated with feudalism. Serfdom was replaced by the supposedly more natural, more highly evolved, system of cash rent and wage labor. Manorial production naturally changed from a subsistence orientation to a commercial one. Commercialization, in turn, naturally stimulated the development of urban production and of long-distance trade.

Notice that the basic causality here is the Weberian idea of rationality: it is rational for humans—or at any rate European humans—to evolve an economy grounded in wage labor and profit. In fact, not only Weber's theory of European rationality but a host of other related theories about Europe's supposed social precocity were inserted at the base of this economic theory of the decline of feudalism and rise of capitalism. For instance, it was claimed by some scholars that the natural development of democracy in the Middle Ages produced, naturally, a political atmosphere that permitted free markets to develop, and with them free labor and free capital. In a word, the standard economic theory of the transformation of feudalism into capitalism, the theory stressing commercialization and trade, is usually a cover for some more basic theory about the inherent rationality and precocity and progressiveness of Europeans.[12]

Brenner puts forward five principal arguments against the trade-based theory of the origins of capitalism. First, he questions why feudal people naturally strive to change from a noncommercial feudal economy to one based on "truck and barter," and then somehow naturally build capitalism out of a commercialized, trade-oriented, postfeudal rural economy. Says Brenner: this implies that people are, somehow, naturally prone toward capitalism, and such an assumption is untenable.[13] (We will see, however, that Brenner himself appeals to a not entirely dissimilar model of human nature in his own theory of the transformation to capitalism.)

The second argument is a faulty one. Brenner asserts that trade during the Middle Ages was a very marginal activity.[14] It consisted, he claims, of insignificant luxury items, for the very small ruling-class population. Quantitatively it could not have been significant enough to have had an impact on rural society. Lords would not, for instance, require very large receipts of cash rent if the only need for cash was to purchase small amounts of luxuries. He argues, in addition, that the towns were in essence parasitic on the countryside, and were not true centers for the production of commodities not produced in the rural areas. Rural–urban trade was not, he believes, quantitatively important, nor did it imply the existence of an important division of labor between urban and rural production systems.[15] Overall, then, trade was not important enough during the Middle Ages to act as a solvent of feudal social relationships and to generate a true market economy and, beyond that, capitalism.

Only one part of this argument is factually correct. As Brenner notes, urbanization during the Middle Ages did not reach the level where it could significantly change rural feudal society. For instance, the old argument that the towns represented a haven for escaped serfs and therefore the growth of towns undermined serfdom by providing a place to which serfs might flee is wrong, because, prior to the end of serfdom in western Europe, towns were too tiny to have played this role to any significant extent. The same objections apply to various theories about the political and social role of towns. For instance, as Brenner notes, they were not, somehow, little enclaves of freedom, of democracy, in the feudal landscape: their social structures were rather rigid, and in any case they were not very hospitable to incoming strangers, like escaped serfs.

But Brenner neglects one very important thing about the towns of Europe. Small though they were throughout the Middle Ages, they were nonetheless little centers of a kind of primitive or incipient capitalism, centers in which some production was organized for profit and some workers were paid wages. Towns, also, were receptacles for the diffusion into Europe of technology, industries, business institutions, and like innovations being developed outside of Europe: they were connected as nodes in a hemispheric trade network. So the towns of medieval Europe eventually became centers of infection, so to speak, for expanding capitalism, *after* the original and crucial causes for the rise of capitalism had taken effect, in a process (and theory) that I have described elsewhere[16] and will summarize later in this chapter—a process involving the inflowing of wealth and the expansion of essentially capitalist enterprise and trade after 1492, after the beginning of colonialism. Thus, the infectiousness of the towns in that later period was not diminished by their relatively small size at the beginning of the period. Brenner rejects the urbanization argument because, for him, both the destruction of feudalism and the rise of capitalism took place not in towns but in the countryside: the first true capitalism, he believes, was an agrarian capitalism.

Was trade really as insignificant as Brenner says it was? Even before the decline of serfdom, the period when peasants had relatively little need for cash, there was vigorous trade in many bulk commodities that were used throughout the peasant economy: salt, iron, utensils, seed, feed, livestock, and many other commodities were traded between villages, and between town and village, and, via the traveling trader and the periodic market, over longer distances. There seems to have been considerable exchange of commodities within the village; presumably some services were purchased with cash. There was demand for commodities from the urban population, the clergy, and other communities and groups. So commerce in this period was hardly insignificant. But after the fourteenth-century crisis and the ensuing decline of serfdom the commercial economy expanded even more, as peasant rents came to be paid in cash and as urbanization increased (a bit). So Brenner is quite wrong to dismiss medieval trade as insignificant, a matter of trivial luxuries. To all of this must be added the fact that trade was even more lively in Mediterranean Europe, a region that Brenner basically ignores.[17]

Brenner asserts next that the growth of trade should not produce a crisis leading to the dissolution of feudalism. The lords would not transform themselves into rural capitalists simply because of the growing opportunities to sell the products from their estates and the growing opportunities to purchase trade items. Nor would they have an incentive to free the serfs simply because commerce was expanding. Brenner is doubtless right in saying that the lords' social status was inseparable from the kinds of services provided them by serfs and tenants, and the growth in commercial opportunities would not lead most of them to alter their values or their way of life. On the other hand, if we introduce into the picture the crisis that took place in the fourteenth century, a crisis in which rural population and the income of the landlord class declined, then Brenner's argument loses force. Lords could not obtain enough labor to maintain their own income level and life style without rather serious adjustment. Given that commerce and trade had expanded, it seems likely that landlords were stimulated to increase production for sale, to reduce costs and regularize output, as a means of maintaining their social system. Moreover, the evidence is compelling that the fourteenth-century crisis led to a transformation of social relations in the western European countryside, with serfdom giving way, broadly, to tenant farming in which rent was demanded in cash.

The fourteenth-century crisis led to the fall of serfdom—not instantaneously, not everywhere, and not entirely—by increasing the bargaining (or class) power of peasants in a situation where peasant labor was scarce. We (and Brenner) are talking about the relative power of two classes, lords and peasants, and the crisis gave the working class greater power to resist the lords' demands that they provide unpaid labor, supply products grown on their holdings, and the like. Peasants gained relative to lords, and this forced the lords to commute labor services and accept payment of rent in cash. But cash rent had meaning only because the overall economy was now commercialized: there was a market for the peasants' produce, and there was a market in which lords could buy things with the cash paid to them as rent. In short, the commercialization of the medieval economy *did* have much to do with the decline of serfdom and indeed of feudalism.

Brenner reinforces his argument with what he calls, rather ingenuously, "comparative analysis," ingenuously because all of the places compared with one another lie within northern (mainly northwestern) Europe, and true comparison would have to look at all possible cases of a given type and so carry one over to Mediterranean Europe, to Asia and Africa, and Brenner does not do this.[18] He points out, correctly, that serfdom actually increased in eastern Europe during the period when trade and commerce were expanding throughout Europe, that is, the sixteenth and seventeenth centuries. He considers this fact to be proof positive that there is no definite connection between the decline of serfdom and the rise of trade: serfdom can decline correlatively with the rise of trade in one region (western Europe) while it increases with the rise of trade in another region (eastern Europe). This argument is easily answered.

The decline of serfdom took place in western Europe during (mainly) the fourteenth century; to compare this with the rise of serfdom two to three hundred years later in a very different region is unfair. In western Europe the rise of commerce took place gradually over the centuries in a region already well developed agriculturally and with a dense population. In eastern Europe the rise of serfdom took place in a very different social and environmental setting. Population was much less dense, and expansion of agriculture to some extent involved a spatial enlargement of the agricultural region. When the increased western European demand for grains in the sixteenth and seventeenth centuries stimulated agriculture in the eastern region, landlords there were confronted with serious labor shortages and found it profitable to intensify the labor services on peasants, enserfing some of them and deepening the burden of serfdom for others. So Brenner cannot properly "compare" these two regions, different in character and at very different times, and claim then that he has proven that the decline of serfdom has nothing to do with the rise of commerce.[19] Indeed, the rise of commerce in the Middle Ages is qualitatively a different thing from the rise of commerce in the early-modern period, after the discovery of America and the explosive expansion of Europe's economy that took place in consequence of that event.

How, then, does Brenner explain the transformation—the fall of feudalism and the rise of capitalism? We can begin by situating Brenner's theory in its geographical and historical perspective.

First: geography. For Brenner, the world outside of Europe was not at all involved in the transformation. Indeed, he derides the idea that we have to look at the non-European world, and attacks writers like Wallerstein and Frank who suggest that we should do so.[20] Brenner says nothing whatever about the nature of medieval societies in Africa and Asia. Clearly, he does not consider it necessary even to raise the matter of comparing European and extra-European societies in order to find out why the former developed capitalism and the latter did not. He does not mention influences on Europe from Africa and Asia. He ignores the late-medieval trade between Europe and these two continents. Thus: the rise of capitalism is a strictly *European* fact.

It is a *northwest European* fact. The transformation did not involve southern or eastern Europe: only the West. Southern Europe is basically not discussed, eastern Europe serves only (as we saw) for "comparison" with the West. But within northwestern Europe, only England is really central to the transformation. France is discussed in some detail, but Brenner does so (as we will notice) in order to show—another "comparison"—why the transformation *did* take place in England and did *not* take place in France. So the decline of feudalism and rise of capitalism is an *English* fact.

There is one further geographical restriction. Towns were not involved in an important way. Brenner believes (contra Sweezy and most historians) that the internal development of European urban life was not important in the rise of capitalism, nor were towns important as havens for escaped serfs, nor was rural–urban trade of significance. The rise of capitalism therefore is a *rural English* fact.

So Brenner's theory has this simple geography: there is distance-decay of interest and relevance as we enlarge the scale, from rural England to England as a whole, to western Europe as a whole, to Europe as a whole, to the world as a whole. The place where feudalism died and capitalism was born was a very small region indeed: rural England.[21]

The geography of Brenner's theory about the origin and rise of capitalism is easily seen when we map it. One simple cartographic method cam be used to examine the spatial patterns associated with this theory. (Basically the same method will be used as part of the analysis, in Chapter 6, of Michael Mann's theory about the westward march of history.) A useful way to get at the geography that is explicit or implicit in many historical texts is to locate and map the dated place-name mentions, or DPMs. A text may, for instance, refer to "Athens." Typically, when the text mentions Athens it does so in a time context that is reasonably clear: say, the fifth century B.C. This yields a dated place-name mention (DPM). All such DPMs can be plotted on maps. We prepare, for a given historical text, maps depicting the world or part of the world in stated periods or historical intervals. We then put dots on each map for all (or a sample) of DPMs that fall within that historical interval. This tells us, rather faithfully, how much attention, or salience, the author gives to the world or various parts of it for a given period in history.[22]

Places, with their proper place-names, are mentioned throughout Brenner's texts, and most of the place-name mentions are either explicitly or inferentially dated: sometimes to the year, sometimes to the century, and sometimes to the historical epoch: medieval or early-modern. Figure 1 presents two maps of dated place-name mentions in Brenner's narrative of the origin and early rise of capitalism. One of the maps, Figure 1a, shows all DPMs associated with the Middle Ages; the other, Figure 1b, shows all DPMs associated with the early-modern period. (DPMs for very large regions, for example, "Europe," "western Europe," are not mapped, but I list the frequency of DPMs for all such regions in the caption to Figure 1.)

Somewhat conventionally, I use 1500 A.D. as the break-point between the two epochs in the maps of Figure 1. It happens, however, that this date is also, very roughly, a break-point in Brenner's theory. As we will see later in this chapter, he argues emphatically that the birth and initial rise of capitalism was strictly an intra-European process, in no significant way affected by the "discoveries," early colonialism, and other extra-European events and influences. Thus, a large part of his argument is designed to show that the crucial processes in the origin and rise of capitalism—and modernity—had taken place before 1500, and so could not possibly have resulted from the events occurring after, and in consequence of, the "discovery" of America in 1492.

All of the DPMs in Figure 1a lie within Europe; in other words, places outside of Europe are not mentioned at all for the medieval period.[23] DPMs for the early-modern period, plotted on Figure 1b, are still clustered in northwestern Europe, but eastern Europe is now given somewhat more emphasis and the extra-European world is mentioned a number of times. These changes in the pattern reflect Brenner's view that capitalism began to diffuse outward to the rest of the world after its birth in northwestern Europe; in part, also, the changes are part of a rebuttal to Frank and Wallerstein, who mention some of these places in eastern Europe, the Caribbean, and so forth, in connection with their supposedly "Third-Worldist" theories about the origins and early rise of capitalism. The two maps taken together show clearly that Brenner's theory is classical Eurocentric tunnel history.

If the geography of Brenner's theory is very clear and definite, its history is both broad and indefinite. The transformation, for Brenner, oc-

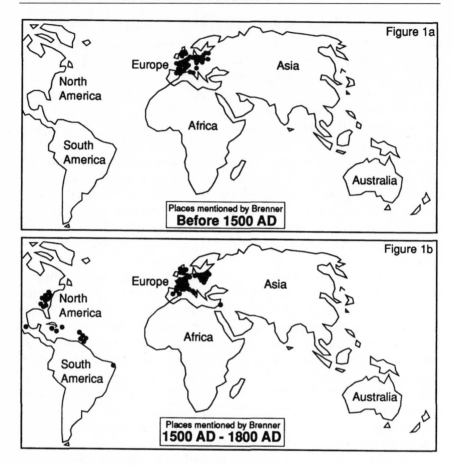

FIGURE 1. Dated place-name mentions (DPMs) in Robert Brenner's narrative of the origin and rise of capitalism. *Figure 1a*: DPMs with dates before 1500 A.D. *Figure 1b*: DPMs for the period 1500–1800 A.D. Not shown on the maps are DPMs for very large regions: "Europe" (42 mentions), "Western Europe" (31), "Eastern Europe" (23), "Continental Europe" (7), "South America" (3), "The East" (3), "North America" (2), "Northeastern Europe" (2), "Africa" (1), "Atlantic World" (1). (The non-European DPMs all refer to the post-1500 period.) The salience of England and France is not adequately revealed by the dot pattern: the place-name "England" appears 92 times ,and places within England are mentioned 21 times; "France" appears 73 times and places within France 48 times; thus French and English place-names account for two-thirds of the 351 DPMs on the two maps. *Sources*: Brenner (1977), (1985a), (1985b).

curred over four centuries. Many would agree that the process was a long and slow transition from fourteenth-century feudalism to eighteenth-century capitalism. Brenner does not argue this way, however. The significant event was the arrival of capitalist agriculture in the English countryside during a rather brief, almost revolutionary, epoch, apparently in the middle or late part of the fifteenth century. But Brenner telescopes a lot of history into this brief period, throwing into it events from the fourteenth, fifteenth, sixteenth, and seventeenth centuries, treating these events as though they were coterminous; even assigning seventeenth-century causes to earlier effects.[24]

We come, then, to the theory itself. Class struggle explains why feudalism collapsed and capitalism rose and triumphed. Throughout the Middle Ages, throughout western Europe, the lords and the peasants were engaged in ceaseless struggle over control of the means of production and surplus labor and product. Although other classes existed in this feudal society, the basic contending classes were landlords and peasants, and in particular lords and serfs. Brenner argues that class struggle in feudalism eventually produced an insurmountable crisis, and feudalism therefore gave way to a new and higher mode of production, capitalism.

But scholars are by no means agreed as to when that crisis occurred and how and why the transformation took place. A major difficulty is posed by the fact that the very serious crisis of the fourteenth century effectively led to the end of classical serfdom (quickly in some areas, slowly in others) but not to the end of feudalism as a general system: the lords remained the ruling class, the peasants were still to some extent under their legal control (for instance, they could not, typically, leave the manor without the lord's permission), and the class struggle between peasants and landlords continued for some time thereafter. Feudalism as a legal and political system ended, formally (in a sense symbolically), in 1688 in Britain, and even later in most other parts of western Europe. Industrial capitalism did not appear, really, until late in the eighteenth century. Even rural capitalism did not really become significant until after 1600. The question, then, is: what happened *after* the fourteenth-century crisis? Why did feudalism still survive? Did it then gradually crumble, without any further crisis, or simply transform itself smoothly, gradually, into capitalism? If the fourteenth-century crisis did not kill feudalism, what did?

Brenner confuses the various phases of the process and so mixes together the events of several centuries, positing that there was a quick revolutionary transformation that occurred roughly in the late fifteenth century, and describing this revolutionary transformation in such a way that

it contains processes that we know were characteristic not of the fifteenth century but of the fourteenth, sixteenth, and seventeenth.[25] It
may seem odd that a historian would make such errors, and to understand
it we must appreciate one additional attribute of Brenner's theory. It is
not entirely an empirical theory; it contains a great deal of mysticism.

Brenner, like some other Marxists, holds to a very mystical conception of capitalism. Capitalism is conceived to be an entity, an essential
thing. When it arrives, it does so complete and entire, as though it were a
god descending from Olympus to govern human affairs. So one does not
really think of a "transition" to capitalism: there comes a kind of mystical
moment when it arrives and takes over. The capitalism that (according
to Brenner) appeared in rural England in the late fifteenth century is the
same essential capitalism that governs England today.[26] Its essence is the
same. Over time, it develops in various ways—for instance, equipping itself with manufacturing industry—but it remains the same entity. And it
retains the same essential properties, some of which are quite mystical.
Capitalism brings with it instantaneous *rationality*. Suddenly technological inventiveness and innovativeness appear; they were not really present
during the feudal age, says Brenner.[27] Suddenly working people are "free,"
that is, they begin to make economically rational decisions in a free labor
market. Suddenly society (English society) acquires an "economy."[28] And
more. This mystical notion of capitalism substitutes for an empirical theory about the transition: the merely empirical facts may suggest a long,
slow transition, with many complex and contradictory happenings, including some regressions toward classic feudalism—no matter. At one
mystical historical moment (or year, or handful of decades) capitalism
appears and transforms rural England.

Here, now, is the core of Brenner's theory. In the fourteenth century,
the English peasants basically won their freedom. The elimination of
serfdom set in motion several processes that then swept away feudalism.
Since peasants were now free, they would tend to rise or fall in economic
status, depending on such matters as the size of their holdings. This was a
process of differentiation of the peasantry into status groups, which eventually became classes. The less successful peasants remained as subsistence farmers or lost their holdings and became landless laborers. The
more successful farmers now negotiated with the landlord/lords to acquire leases on fairly large holdings, holdings now large enough so that,
with hired labor, they could produce a profit and favor the accumulation
of capital.[29] But the key factors in the process were these: First, the elimination of serfdom freed the minds of the peasants so that they could begin, rationally, to think up ways to improve agricultural production. Sec-

ond, the new freedom from serfdom meant that agricultural laborers would move around in a labor market, taking work where the compensation was highest.[30] These were the two essential features of capitalism: capitalist rationality, leading to technological innovation; free wage labor, leading to efforts to reduce the cost and raise the efficiency of labor. The larger peasants now became small businessmen, leasing land from the lords, hiring labor, competing with one another, and accumulating capital or—if unsuccessful—going out of business. In short, the standard menu of attributes for a modern small business enterprise.

There are a number of very large problems with this theory. The largest is a matter of timing. Serfdom basically ended in the fourteenth century, but capitalist agriculture—the model we have just looked at—was almost unknown prior to about 1520 and did not really become widespread for the next hundred years or more. Indeed, when Brenner goes into detail about the supposed attributes of capitalist agriculture in rural England, he tends, more often than not, to be describing the kinds of farms that were characteristic of the eighteenth century—some 400 years after the collapse of serfdom. The error is most glaring when he waxes enthusiastic about the new rationality and the technological innovations that it produced. He speaks of "revolutionary innovations," but there were none such—until centuries later.[31] Brenner bases his argument here mainly on one authority (Kerridge); many others deny that there was such a revolution, at least in matters of agricultural technique, some maintaining that the real revolutionary epoch was the eighteenth century, or maintaining even that nothing truly revolutionary occurred in the English rural landscape before the industrial revolution.[32]

Let us unpack this problem into smaller and more manageable ones. First, Kerridge's sixteenth-century agricultural revolution is still, as to timing, a century too late to satisfy Brenner's theory: serfdom gave way to "free" tenantry before the end of the fourteenth century. Second, Brenner argues that technological advances, after the freeing of the peasants, led to the enlargement of holdings and the creation of capitalist farms. But the technological advances that had this effect occurred a couple of centuries later: effect therefore precedes cause. The two revolutionary technological advances actually discussed by Brenner were not at all revolutionary. He cites an innovation in irrigation technology (floating of water meadows), but this was neither very innovative (irrigation being an old art) nor very important.[33] Brenner next cites a new (for England) system of rotation involving the alternating of improved pasture with cropland ("convertible husbandry"), but this did not intensify production (some other, older, rotations were very much more intensive) although it was a solid advance in pasture technology; and, rather then being a revolutionary advance, it was used in Flanders—not very far away— in the early fourteenth century.[34] So the entire argument about a sort of instantaneous appearance of "rationality," and then, immediately and directly, the beginnings of revolutionary technological advance, is simply empty.

Brenner seems to have in mind the marvelously rapid technological advance that accompanied capitalism during and after the industrial revolution, the process that Marx identified as a central feature of *modern* capitalism, new technology being a crucial strategy for firms in their competition with other firms. Brenner (quoting Marx) insists that the "rational" process of constantly revolutionizing technology is an essential attribute of all capitalism, then casts all of this back into a time when, in fact, constant revolutionary technological advance just did not take place. The mysticism of his concept of capitalism overrides the facts, and the eighteenth century is pushed back to the fifteenth century.

Next there is a serious difficulty with Brenner's conception of medieval technology. He holds a very contradictory image of the peasantry. He thinks that medieval peasants were not at all innovative as to technology but that some peasants became marvelously innovative as soon as they were touched by the magic wand of capitalism. The reasoning here is that peasants are conservative and unchanging, traditional, so long as they own the means of production, the land, and gain their livelihood from it.[35] If they are serfs, says Brenner, they have no incentive to think up and introduce technological advances since the lords will reap the benefit. But they become innovative and progressive when land is a commodity and they must produce for sale in order to be able to pay the rent on their land. Unaccountably, Brenner seems to think that serfs owned the means of production (and throughout his argument he describes peasants, medieval and modern, as though they were owners of land whereas in fact most were not proprietors at any time in this period, in England or elsewhere in northwestern Europe).[36] This error aside, the fact is that peasants were not hidebound and traditional. We can infer this from modern research that disproves the contemptuous attitude that European "modernization" theorists hold about peasants and their supposed irrationality, traditionalism, and the like—an attitude that Brenner clearly shares. We know this also from the research, which has uncovered a broad array of peasant-generated technological advances in the Middle Ages.

Brenner makes a somewhat similar, and equally fallacious, argument about the feudal landlords who, he says, have no incentive to innovate technologically because they are, in essence, satisfied with their social situation.[37] Brenner makes two errors here. First, there were periodic crises throughout the Middle Ages, and landlords were very frequently faced with a lack of delivery of surplus and a need to increase it. Brenner imagines that the standard way to do so was to squeeze the peasantry ever more tightly rather than to attempt to improve production methods, on the demesne farm or on the peasant farms.[38] Granted that feudal lords were expert squeezers, nonetheless many of the estates, lay and ecclesiastical, made serious and important efforts to improve agricultural methods and introduce better technology. Brenner's next error is to assume that squeezing had no limit: nowhere does he notice that medieval serfs were suffering exploitation to and sometimes beyond the subsistence line. Another fact about the squeezing process that Brenner ignores: when peasants were forced to increase the delivery of surplus produce, they were under intense pressure to increase their levels of production, hence were likely to (and often did) innovate, technologically and in other ways, in order to increase production. In other words, it is simply untrue that serfs (and also landlords) had no incentive to innovate. Second, Brenner dismisses one of the common arguments of Marxist theory: in all class societies, the ruling class is always seeking to increase its wealth; it is never satisfied, and the system is never in equilibrium. Brenner conceives feudal society as having been governed by completely different rules than those that apply to capitalist society. It did not have an "economy." It had relatively little exploitation. And the feudal ruling class supposedly was satisfied with a certain level of income from its peasantry, or enough so as to be willing merely to squeeze the peasants as far as possible and not to attempt to increase total production. Overall, says Brenner, this society was "stagnant."[39]

We come now to what is probably Brenner's strangest proposition. His theory is self-consciously Marxist and self-consciously grounded in class struggle. In Marxist theory, class struggle tends to produce advances in cultural evolution because, putting the matter simply, the exploiters lose. For Brenner, the ruling class was defeated to the extent that peasants secured their freedom from serfdom. But this did not bring about the collapse of feudalism as a mode of production. That occurred (in England) roughly one hundred years later, according to Brenner, and it occurred because the ruling class *won* the class struggle. Brenner argues that, if the peasants had really won in the fourteenth century, the result would have been, not rural capitalism, but a society of freeholding peasant proprietors. Because peasant proprietors (in Brenner's thinking) are not innovative, are satisfied to have a bucolic existence on their subsis-

tence holdings, this form of society would not have gone through a transformation to capitalism.

Brenner now points to France and makes one of his limited (and invalid) comparisons. In France, he says (inaccurately), the peasants won
definitively, so freeholding peasants really came to dominate the society,
established cozy links with the crown against the landlords, and as a result managed to maintain their position.[40] This explains why capitalism
did *not* arise in France. In England, on the other hand, the peasants *lost*.
They secured the ending of serfdom, but they did not succeed in winning
full proprietorship of their land: they remained tenants of the same landlords.[41] As a result, says Brenner, there appeared a subclass of peasants
who parlayed tenancy into capitalist agriculture. They negotiated rents
with the landlords, rented larger and larger holdings, hired labor, and so
became capitalist farmers, paying a portion of their profits to the landlords, just as modern small businesses pay rent to the owners of their fac-
tories and offices. For Brenner, this was the real cookpot of capitalism.[42]
So the fact that English peasants *lost* their class struggle is the crucial explanation for the ending of feudalism and the rising of capitalism. This
turns the class-struggle theory on its head.

NEO-WEBERIAN EURO-MARXISM

Brenner wants to label Frank, Wallerstein, and company "Neo-Smithian
Marxists" because, in essence, they pay so much attention to commerce,
to trade, to urbanization. He never quite demotes them from the status of
"Marxist," but this is implicit in his argument, as commentators have
pointed out.[43] Of course, Marx did not neglect commerce, trade, and urbanization. But Marx gave ultimate causal authority to class struggle.
Brenner claims to do so, too. For him, Sweezy, Frank, and Wallerstein
abandon class struggle in favor of commerce and the rest. This is quite
unfair to Sweezy, who keeps an eye on urban class struggle, and also to
Frank and Wallerstein, who notice that there was class struggle in Latin
America as well as Europe. But this is beside my present point. Brenner's
theory actually gives a rather minor role to class struggle.

Like everyone else who writes about the medieval origins of capitalism, Brenner reports the class struggle between serfs and lords and regis-
ters the fact that a change in class structure was under way, but he hardly
goes farther. The essence of his theory is the argument that a commerce-
minded accumulating class of tenant farmers rose in England, in the
midst of all such changes. These yeoman-tenants, he says, were a kind of

class product of prior class struggle, along with class differentiation. But that prior class struggle—mainly between serfs and landowners—took place in many places, not only rural England. Brenner's key proposition is that English tenant farmers developed into capitalist farmers—the first capitalists—not because they struggled with anybody but because they had not been able to gain ownership of the land. This, for Brenner, implies that they did not fall prey to the noninnovative conservatism that (he thinks) afflicts landowning peasants. Since they did not own the land and were not serfs, they were "free" to accumulate. (Landlords were, in this conception, "unfree."[44]) All of this establishes the possibility, feasibility, desirability of capitalist activity "at the highest level of technology."[45] And Brenner now turns to a very unmarxian theory to explain—really explain—the rise of capitalism. This is the idea of technological rationality, which, for Brenner, emerges in this one unique class and place (here ignoring the existence of cash tenancy and agricultural wage-labor in other regions and continents).

This view is less Marxist than Weberian, and Brenner's essential theory is really closer to Weber's than it is to Marx's. Brenner differs from Weber in believing that this rationality is not a permanent attribute of European people but rather it descended on Europe all at once (so to speak), arrived rather suddenly, fully formed, at the magical moment when the class of (commerce-minded) English yeoman-tenant-farmers began to accumulate. Yet this is not so far from Weber, who also fixates on the importance of capitalist rationality and its supposed burgeoning at a magical moment (in the Reformation). So long as people are being called names, let's call Brenner a "Neo-Weberian Euro-Marxist." Just in fun.

OTHER VIEWS

Name-calling is not entirely irrelevant to Brenner's case, since, if "Third-Worldists" are banished to "Neo-Smithian" territory, then the "Marxist" terrain would seem to have been secured for Brenner. But it has not been. For one thing, Marxists have been debating for a long time, and still are debating, the question whether urban or rural forces were most crucial in bringing about the transition from feudalism to capitalism. Brenner does not try to mislead anyone about the fact that he is continuing the famous Dobb–Sweezy debate,[46] advancing a basically (or neo-) Dobbist case against Sweezyism, and against all the rest of the Marxists who think that the rise of urban processes was as consequential, or more so, than the contemporary decay of feudalism in the countryside. But many readers of Brenner have plain forgotten the older debates (which go back to Marx's own uncertainty on this matter) and have decided that to support Brenner's conception of the origins of capitalism is to defend all of Marxism against "Third-Worldism."

The debate is partly a matter of *place*. Is Brenner right to locate the causation in the lands worked by tenant farmers, not landlords?[47] In rural England, not in urban England? In England and not also in France?[48] In western Europe but not also eastern Europe?[49] Not also southern Europe?[50] Not also Asia?[51] Not also Africa?[52]

All of the important attributes of late-medieval English rural society, including its class struggles, were to be found during the same period in many other places. This includes untied peasantry, cash tenancy, rural wage-labor, large-scale production for sale, peasant struggle, and much more—not excluding a certain pace of agricultural innovation. The evidence demonstrating all of this is fairly abundant.[53] Likewise, it is clear that urban and peri-urban processes, including large-scale material production for commerce, were no less advanced in southern Europe and Asia than in northern Europe during this period.[54] What is not clear is the relative importance of decay or decline of a feudal or tributary mode of production—class-based agricultural production in landlord-dominated regions—as against a development of urban production, urban class processes, and a class of landlords beginning to invest in non-agricultural enterprises at home and overseas. I suspect that both were aspects of a single historical process. Probably there is no contradiction between Dobb and Sweezy at a world-system level of analysis.

For these reasons, I argue that northwestern Europe was not in any sense more developed, or more pregnant with development possibilities, than many other regions prior to 1492. The crisis of the late-medieval feudal or tributary economy seems to have been occurring in many regions. The processes singled out by Brenner for northwestern Europe were going on elsewhere. (I developed this argument in Volume 1.)

Since the great majority of the producing class were peasants, it is certainly true that rural exploitation and rural class struggle were at, or close to, the heart of the matter. But the complex of urban and peri-urban facts and processes, including an emerging urban bourgeoisie and urban working class, merchant activities, manufacturing activities, large-scale long-distance commodity movement, business institutions, and a variable but always significant degree of political autonomy were also very important in their own right: the town was not simply a dependency

of the countryside, as Brenner impli
emerging crisis in rural feudalism.

The urban and peri-urban pheno
northern Europe before the High Middle
importance in southern Europe and par
early date. The late-medieval urban compl
production that is best thought of as incipie
or protocapitalism—it is not a form of feudali
modity production"—was found in many par
sphere. Many extra-European cities were as fully
talist production, capitalist class relations, and capi
most advanced cities of Europe, but all were conne
circuit of crisscross diffusion, extending throughout t

But also very widespread across the hemisphere was
duction involving rural wage-workers in the production
sometimes in peri-urban zones and larger hinterlands of cit
in purely rural regions. In all of this there is a double-barrele
Brenner's position. Brenner is mistaken in searching for fund
ral transformation only in northern Europe: he should searc
Fujian, Vijayanagar, Kilwa, the Nile valley, the upper Niger val
so on. But Brenner is mistaken also in dismissing urban transforma
the later Middle Ages, transformations of class, production, and r
more, transformations that were going on at an impressive rate in sou
ern Europe, Asia, and Africa.

A number of writers, Marxists and non-Marxists, have argued the
broad position that I have just summarized: that processes of change out
of something like a feudal or tributary mode of production and toward
something like capitalism were occurring in many world regions during
the late Middle Ages—not merely in England and the Low Countries—
and were taking place in urban as well as rural regions. My own view, not
very different from the views of Andre Gunder Frank, Janet Abu-
Lughod, and Samir Amin,[56] is that the transition was at about the same
rate and level in all three continents in 1492.

If, indeed, the processes of historical change out of feudalism and toward capitalism (or something like capitalism) were going on in various
parts of the Eastern Hemisphere in the late Middle Ages, and northwestern Europe was in no sense a leader, how do we explain the fact that Europe rose, Africa and Asia did not, and northwestern Europe developed
industrial capitalism and empire? My own view focuses, again, on the
matter of place: of location, or accessibility.) We start with a conception

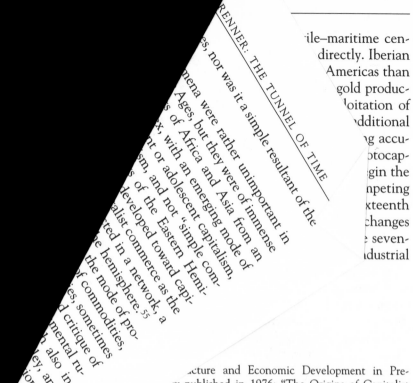

·ile–maritime cen-
directly. Iberian
Americas than
gold produc-
loitation of
additional
ng accu-
otocap-
gin the
npeting
xteenth
changes
e seven-
dustrial

...cture and Economic Development in Pre-
...ly published in 1976; "The Origins of Capitalist
...o-Smithian Marxism" (1977); "The Agrarian Roots of
...o), originally published in *Past and Present* (1982).

...lpin, eds. *The Brenner Debate: Agrarian Class Structure and Eco-
m Pre-Industrial Europe (1985).

...or instance, Torras, "Class Struggle in Catalonia" (1980); Hoyle, "Tenure
...nd Market in Early Modern England: Or a Late Contribution to the Brenner
...e" (1990); Katz, "Karl Marx on the Transition from Feudalism to Capitalism"
(1993); Overton, *Agricultural Revolution in England* (1996).

5. Frank's present views are not *dependentista*. See his *ReORIENT* (1998).

6. M. Cooper, "Town and Country in the Great Transition: Adam Smith et al. on Feudalism and Capitalism" (1980), pp. 81–82.

7. Browett, "Into the Cul de Sac of the Dependency Paradigm with A. G. Frank" (1980), p. 111.

8. Macfarlane, "The Cradle of Capitalism: The Case of England" (1988), p. 191.

9. See, for example, Corbridge, *Capitalist World Development: A Critique of Radical Development Geography* (1986), pp. 38–44; O'Brien, "Path Dependency, Or Why Britain Became an Industrialized and Urbanized Economy Long Before France" (1996).

10. For a much later period (seventeenth and eighteenth centuries) and in a different context (Brenner's argument that underdevelopment was not inevitable for the colonial world), brief mention is made of Barbados and colonial North America in Brenner, "Origins".

11. In Brenner, "Agrarian Roots": "Nor can there be any question but that the Malthusian model, in its own terms, has a certain compelling logic. Malthus's model was cor-

rect not for the emergent industrial economy he was analyzing, but for the stagnant back-ward society from which this had arisen" (p. 14). "In terms of its special premises and the small number of variables it entails, secular Malthusianism seems almost foolproof" (p. 15). "Malthusian theorists have indeed isolated an important pattern of long-term eco-nomic development and stability" (p. 18). "The [fourteenth-century] crisis of productivity led to demographic crisis, pushing the population over the edge of subsistence" (p. 33). "[Marked] demographic decline . . . came to dominate most of Europe . . . the famous Malthusian 'phase B' " (p. 51). "[The] rising population . . . [and prices] of the 16th and 17th centuries [led] merely to a renewal of the old Malthusian cycle of underdevelop-ment" (p. 61). In Brenner, "Agrarian Class Structure": "My intention [in the 1976 article] was not to deny the existence of [the Malthusian] two-phase cycles; it was to expose the limitations of the neo-Malthusian cum Ricardian models . . . in actually *explaining* the long-term patterns" (p. 217). "No one would deny that continuing demographic increase in the face of declining labour productivity sooner or later leads to an imbalance between population and resources—ultimately to poverty, famine and death" (p. 223). "[The] de-mographic interpreters . . . [cannot] tell us why their Malthusian premise of the non-development of the productive forces essentially held true throughout a whole epoch, but then ceased to do so" (pp. 224–225). Political centralization "disrupted the 'normal' Mal-thusian mechanism for bringing population into line with production" (pp. 241–242). "Continuing population growth had *eventually* to result in widespread poverty and fam-ine" (p. 265). "The acceleration of political centralization . . . short-circuited the needed Malthusian adjustment" (p. 270). "Dutch agriculture experienced no tendency towards a demographically powered evolution . . . the familiar Malthusian pattern" (p. 320). The English seventeenth-century economy had the "capacity to maintain demographic in-crease beyond the old Malthusian limits" (p. 325). Also see Brenner, "Agrarian Roots," pp. 27 and 46; Brenner, "Agrarian Class Structure," pp. 227, 230, 236, 269; Brenner, "Or-igins," p. 35.

12. Among these scholars are Eric Jones, John Hall, and David Landes, whose views we discuss in other chapters of this book.

13. Brenner, "Agrarian Roots," p. 17; "Origins," pp. 28–29, 40, 42, 45–49, 61, 83.

14. Brenner, "Agrarian Roots," p. 25; "Agrarian Class Structure," pp. 241, 274; "Origins," p. 47.

15. Brenner, "Agrarian Roots," pp. 38–39; "Origins," p. 47.

16. Blaut, "Where Was Capitalism Born?"; *Colonizer's Model*, Volume 1.

17. See the critique of this discussion in Torras, "Catalonia." On towns and trade, see, for example, Hilton, *English and French Towns in Feudal Society: A Comparative Study* (1992).

18. See Brenner, "Agrarian Roots," pp. 21, 30; "Agrarian Class Structure," pp. 220, 221, 260.

19. See, in criticism of Brenner on this matter: in Aston and Philpin, *Brenner De-bate*: Wunder, "Peasant Organization and Class Conflict in Eastern and Western Ger-many" (1985), pp. 91–100; and Le Roy Ladurie, "A reply to Robert Brenner" (1985), pp. 101–106. Also see Dodgshon, *The European Past: Social Evolution and Spatial Order* (1987), p. 12.

20. Brenner, "Origins," pp. 27–33, 38–41, 53–63, 67–68, 82–92.

21. It should not be supposed that Brenner's extreme emphasis on European places simply reflects the fact that his two *Past and Present* articles have "Europe" in their titles. The titles and the contents of these essays reflect the fact that Brenner is theorizing that capitalism and modernity arose in Europe and in Europe alone. In Brenner's 1977 *New*

Left Review essay, which bears the more revealing title "The Origins of Capitalist Development," there is simply no mention of non-European places for the period prior to the seventeenth century. Europe, with its "historically developed class structures," has been and remains the center of development (p. 91). "[The] dynamic of capitalist development [is] in a self-expanding process of capital accumulation by way of innovation in the core" (p. 29)—the core being, historically, the European world.

22. This procedure is discussed in Blaut, "Mapping the March of History" (1993b), where I describe the method and apply it to several world history texts and historical atlases. This essay will be incorporated in revised form in Volume 3 of *The Colonizer's Model of the World.*

23. French place-names are mentioned about as frequently as English place-names because much of Brenner's narrative is an argument that capitalism could not have arisen in France. I discuss this "comparison" between England and France later in this chapter.

24. Brenner, "Agrarian Roots," pp. 35, 46–54; "Agrarian Class Structure," pp. 215, 252, 274, 276, 282, 294–297, 200–206, 309–311, 314–316.

25. I base this on a careful examination of the Brenner texts in Aston and Philpin, *Brenner Debate*, noting, for each statement of historical fact, the explicit or implicit date assigned to that fact.

26. Brenner, "Agrarian Class Structure," pp. 234–236, 275, 293, 295–301; "Origins," pp. 30–32, 61, 67–68, 78.

27. Brenner, "Agrarian Roots," pp. 29, 31, 33, 50, 59, 63n; "Agrarian Class Structure," pp. 214, 233–236, 290, 303, 206–316; "Origins," pp. 26, 42–43, 46, 68.

28. Brenner, "Agrarian Class Structure," pp. 227, 228n, 233; "Origins," p. 30, 37. Many Eurocentric Marxists share this quaint view that feudalism did not truly possess an economy—that exploitation was somehow carried out under the aegis of politics and religion but not economics.

29. Brenner, "Agrarian Roots," pp. 30–63; "Agrarian Class Structure," pp. 274–327.

30. Brenner, "Agrarian Class Structure", p. 33.

31. Brenner, "Agrarian Class Structure", pp. 32–33, 46, 49–51, "Agrarian Roots," 214–215, 296–297, 306, 308–311, 315–316. See the critique by J. Cooper, "In Search of Agrarian Capitalism" (1985), pp. 141–143.

32. Eric Kerridge, *The Agricultural Revolution* (1967). For opposing views, see, for example, Titow, *English Rural Society, 1200–1350* (1969). For critiques of Brenner's position, see, for example, Overton, *Agricultural Revolution,*; Croot and Parker, "Agrarian Class Structure and the Development of Capitalism: France and England Compared" (1985), pp. 79–90; J. Cooper, "In Search." Some economic historians believe that there was *no* agricultural revolution during the Middle Ages and down to the period of the industrial revolution, when industrialization, urbanization, and population growth in England correlated in space-time with increases in agricultural productivity and the introduction of truly novel farming technology—notably new fertilizers and much-improved steel implements—leaving open the question: which caused which? See Clark, "Renting the Revolution" (1998) and "What Was Real Agricultural Output in England in 1700 Compared to 1850 or 1860?" (1999); Grantham, "Contra Ricardo: On the Macroeconomics of Pre-Industrial Economies" (1999). Many economic historians accept the concept of an agricultural revolution in England but date it long after the period claimed for it by Kerridge and Brenner; see, for example, Overton, *Agricultural Revolution*; Beckett, *The Agricultural Revolution* (1990).

33. See Brenner, "Agrarian Class Structure," p. 309; Brenner, "Origins," p. 42; Croot and Parker, "France and England," pp. 79–90.

34. Brenner, "Agrarian Class Structure," pp. 309, 316; Brenner, "Origins," p. 42; also see Croot and Parker, "France and England," p. 80.

35. Brenner, "Agrarian Roots," pp. 59, 63n; "Agrarian Class Structure," pp. 234–236, 306, 311, 313,; "Origins," pp. 36, 43.

36. Brenner, "Agrarian Roots," pp. 48–49; "Agrarian Class Structure," pp. 228–230, 243–244, 265, 306. For the opposing view of landownership in this period, see Cooper (1985), pp. 151–152 and elsewhere; Wunder, "Peasant Organization," pp. 91–100. Also see Hoyle, "Tenure" (1990), and Overton, *Agricultural Revolution.*

37. Brenner, "Agrarian Roots," pp. 31, 43–44. In the late Middle Ages some landlords invested heavily in novel nonagricultural enterprises, including overseas adventures. See Cain and Hopkins, "Gentlemanly Capitalism and British Expansion Overseas: Part I. The Old Colonial System, 1688–1850" (1986).

38. Brenner, "Agrarian Class Structure," pp. 234–235 and elsewhere.

39. Brenner, "Agrarian Roots," p. 41.

40. "Agrarian Roots," pp. 29, 54–55, 59–61; Brenner, "Agrarian Class Structure," pp. 290, 307–311. For a critique, see Croot and Parker, "France and England"; Bois, "Against the Neo-Malthusian Orthodoxy" (1985), pp. 107–118; and J. Cooper, *In Search,* p. 185n.

41. Brenner, "Agrarian Roots," pp. 48–49, 61; "Agrarian Class Structure," pp. 307–311. See Grantham, "Agricultural Supply During the Industrial Revolution: French Evidence and European Implications" (1989), and his "Divisions of Labour: Agricultural Productivity and Occupational Specialization in Pre-Industrial France" (1993).

42. Brenner, "Agrarian Roots," pp. 49–53; "Agrarian Class Structure," pp. 293, 296–302, 315–318. See J. Cooper, "In Search" ("Brenner Sounds like a Tory Defender of the Corn Laws," p. 189).

43. See, for example, Macfarlane, "Cradle," p. 191; Taylor, "The World-Systems Project" (1989), pp. 342–343.

44. See Hoyle, "Tenure," on accumulation by landlords.

45. Brenner, "Origins," p. 32.

46. Hilton, ed., *The Transition from Feudalism to Capitalism* (1976).

47. Hoyle, "Tenure."

48. Bois, "Neo-Malthusian Orthodoxy."

49. Wunder, "Peasant Organization"; Wallerstein, *The Modern World System,* Vol. 2 (1980).

50. J. Torras, "Catalonia."

51. Abu-Lughod, *Before European Hegemony: The World System* A.D. *1250–1350* (1989).

52. In *The Colonizer's Model of the World* I raise (but do not try to answer) the question whether several African civilizations in 1500 were as far behind European and Asian civilizations as contemporary scholarship argues.

53. See Blaut, "Where Was Capitalism Born?" (1976) and Volume 1.

54. See, for instance, Abu-Lughod, *Before European Hegemony;* Frank and Gills, "The Five Thousand Year World System: An Interdisciplinary Introduction" (1992); Frank, *ReORIENT.*

55. Blaut, "Where Was Capitalism Born?"; Abu-Lughod, *Before European Hegemony.* I am not arguing that the emerging system was moving, teleologically, toward the kind of capitalism that later emerged in western Europe. The trend, which I believe was

evolutionary, sharply changed direction after the conquest of America. Nor am I dismissing as unimportant the role played by commerce-minded landlords: see Cain and Hopkins, "Gentlemanly Capitalism."

56. Frank, "Fourteen Ninety-Two Once Again"; Abu-Lughod, *Before European Hegemony*; and Amin, *Eurocentrism* (1989).

57. West African urban centers in 1492 were located well inland; were oriented to land commerce, hence had no major maritime orientation. All other centers that had such a maritime orientation—principally those of East Africa and Asia—were much farther from the Western Hemisphere than were the Iberian centers. See the discussion of this matter in Volume 1.

CHAPTER 5

Eric L. Jones:
The European Miracle

E urocentric history was not seriously challenged until the end of World War II, the period of decolonization, the period when, in Amilcar Cabral's famous ironic comment, colonial peoples "re-entered history." Now scholars from the Third World, and a few from the European world, began to question most of the basic tenets of Eurocentric history. The influence of this critique grew rapidly, and, not surprisingly, the traditionalist Eurocentric historians responded with counterarguments in books that (again not surprisingly) gained much publicity and wide circulation.

Some of these books were panegyrics, announcing in triumphant tones the past and present superiority of Europeans and the beneficial effects that Europe had on the rest of the world during and after the period of colonialism. The most famous of these was W. W. Rostow's 1960 book *The Stages of Economic Growth*, which provided an overview of world history from a triumphalist Eurocentric perspective.[1] At the same time, some historical theorists were trying to reformulate the traditional doctrine to strengthen it and fill its lacunae. Some of these theorists, among them Lynn White, Michael Mann, and John Hall (see Chapters 2, 6, and 7), developed partly new Weberian theories about the uniquely progressive, inventive European mind. Others, mostly mainstream economic historians like Douglass North and Marxist historians like Robert Brenner (Chapter 4) and Eric Hobsbawm, put forth historical theories dwelling less on the idea of European rationality and more on concrete social or economic conditions in historical Europe that supposedly led to Europe's unique "rise." The difference between these two groups of Eurocentric historians was, however, blurred: each made use of the other's arguments to one extent or another.

Eric L. Jones's 1981 book *The European Miracle*[2] was the first systematic attempt to assemble all of these newer arguments for Europe's past and present superiority, along with the traditional arguments, into a comprehensive statement of all the reasons why Europeans have always been superior to everyone else and still are so today. It is certain that this book marked a watershed in modern Eurocentric history writing. Its influence has been immense. Until 1998, the year when David Landes's book *The Wealth and Poverty of Nations: Why Some Are So Rich and Some So Poor*—a more elaborate and detailed presentation of an equally extreme Eurocentric world history—was published, Jones's *The European Miracle* was truly the canonical work. *The European Miracle* is so important a text for any general critique of Eurocentric historiography that I will discuss it in detail here, giving it more attention than I do other works, earlier and later.

The European Miracle is informally written and has the quality of stating most of the Eurocentric positions as flat and undisputed facts, beguiling the unwary reader into thinking that there is little or no contradicting evidence and little or no dispute on these matters in the scholarly literature. Again we should not be surprised to find that this book quickly achieved its canonical status: it went through several printings and two editions, it was adopted widely in university courses, and, perhaps most intriguing if one is thinking about the social history of ideas, it was praised by historians who were well aware of its errors and oversimplifications but clearly considered its message to be so important that these failings were to be excused or ignored.[3]

There were, of course, some polite criticisms from the academic community: he goes too far, especially in his extremely and embarrassingly negative statements about Asians, their character and history; he tries to make history seem to be a science, and overemphasizes the hard factors, such as economics and the environment, and underemphasizes the cultural factors; and so on. Jones responded to these criticisms in a shorter follow-up volume called *Growth Recurring: Economic Change in World History*. In this later (1988) book he briefly restated his basic argument but with some striking changes. *The European Miracle*, first of all, had used extremely insulting language in its descriptions of non-Europeans, language that could be interpreted as racist. Notable examples: Asians. historically, were servile, lazy, and uncreative (pp. 161–167, 231),[4] and seemingly, for Asians; "copulation was preferred over commodities" (p. 15). Africa had no real influence on history except as a source of slaves (p. 153), and Africans in the historic period were mostly hunters, "part of the ecosystem . . . not above it" (p 154). Europe, by contrast, was "a mutant civilization" (p. 45), "a peculiarly inventive society" (p. 227). In *Growth Recurring* there is very little of this sort of language, and there is, indeed, a somewhat more sympathetic and informed view of Asians if not of Africans. Secondly, Jones rethought some—by no means all—of his arguments for European historical superiority; clearly, in the seven-year interval between the publication of the two books Jones did some reading that he should have done before publishing the first book, and he reversed himself on a number of issues.[5] Third, he made it very clear that his arguments are not racist. And fourth, he restated some of his arguments in probabilistic, not absolute, terms, and mentioned exceptions to several of his earlier generalizations: most consequentially, he backtracked from the idea that Europe had been the *only* truly progressive society in history, arguing instead that economic growth of the really progressive sort had indeed taken place for a few centuries in Song China and briefly in Japan. What Europe had accomplished uniquely was not economic growth but "growth recurring."

In this chapter I will dissect the argument of *The European Miracle*, showing it to be false from start to finish, and much more briefly analyze the argument of *Growth Recurring*. I decided to proceed in this way because *The European Miracle* is extremely influential, and still, today, is more widely cited and used than its sequel. An analysis of the arguments offered in *The European Miracle* is an important task for any critique of Eurocentrism in contemporary historiography.

DEVICES

The European Miracle is a pure example of Eurocentric history, and an especially sophisticated one. It advances the unqualified argument that Europe has at all times in history been superior to all other civilizations in all dimensions of culture relevant to economic development, to progress, to modernization. Its sophistication lies in the devices that Jones employs to make his case appear coldly factual, empirical, scientific, scholarly, and therefore convincing to critical readers. I use the loaded word "devices" instead of "approach" or "procedures" because what Jones has to say in this book is not factual, empirical, scientific, scholarly: it is made to appear so by the form of argument. This seems to be the familiar phenomenon of a Eurocentric historian so utterly convinced of the superiority of Europe that he abandons the canons of scholarship and lays out evidence less to demonstrate than to illustrate what is to him obvious.

The form of argument in *The European Miracle* includes the following devices, among others: First, in explaining the superiority of Europe and Europeans, Jones lays great stress on what seem to be solid, material causes, matters of the natural environment, human nutrition and disease, demography, and the like. The psychological and cultural superiority of Europeans is asserted throughout the book, but Jones (unlike Weber) introduces such things as causal factors only where absolutely necessary and only with great subtlety. Jones's second device is to build support for his various arguments by amassing just about all of the traditional beliefs about European superiority, excepting only the beliefs that have been absolutely disproved, and asserting them without any discussion, as though they were established truths, not unfounded and often biased assertions. The mere quantity of these assertions seems, then, to convey the idea that overwhelming evidence exists to support Jones's argument for European superiority. The third device is an appearance of cross-cultural comparative analysis. In *The European Miracle* Jones devotes as much attention to arguments for the inferiority of Asian civilizations as he does to the superiority of Europe. He describes this as a comparison among civilizations. In fact, though, his arguments for the inferiority of Asian civilizations—Africa is dismissed in four pages of negative comments—generally use the same device of listing great numbers of dubious and false traditional beliefs about Asians and Asian societies, giving the false impression that masses of evidence exist to support his arguments, whereas little or none of what he offers is evidence. The fourth device is a claim that an assertion has been "shown" to be true, has been "established," followed by at most one citation to an obscure and sometimes worthless source, as a means of making it seem that what Jones is saying is truly established mainstream scientific truth. The fifth device is the familiar one of telescoping history in the characteristic Eurocentric way. In some cases, modern Europe—already rich, industrial, developed—is compared to ancient Asia, and the superiority of Europe is thus affirmed. In other cases, the poverty and degradation of Asian societies under colonialism is exhibited as though it were the permanent character of these societies, so that Jones can then argue that historical Europe at any given period was superior to an Asia that did not in fact exist at the time in question.

The argument of *The European Miracle* is presented in a straightforward, systematic way. The first part of the book is a listing of all of the supposed reasons why Europe has been superior throughout history. The second part of the book is a listing of all of the supposed reasons why Asia has been inferior. Let us now run down these two lists, roughly in the order in which they are set forth by Jones, and refute both myths: that Europe was superior; that Asia, and non-Europe as a whole, was inferior.

"THE QUALITY OF EUROPEANNESS"

The book begins by summing up what Jones calls "the quality of Europeanness" in these words:

> Europe did not spend the gifts of its environment "as rapidly as it got them in a mere insensate multiplication of the common life." (p. 3)

Here we have three key propositions, which are developed throughout *The European Miracle*: (1) Europe had unique environmental "gifts"—in other words, Europe's physical environment gave it superiority. (2) Non-Europeans wasted their resources on population growth rather than development; they "spent" these resources on mere "multiplication of the common life." (3) To allow uncontrolled population growth in this way, thus wasting resources and losing the chance to develop economically, is "insensate": non-Europeans, in a word, are irrational. Jones makes many other assertions in the course of this book, but these three—Europe's environmental superiority, intellectual superiority, and freedom from Malthusian population disasters—are probably the most crucial generalizations.

Next we have a series of assertions that are designed to establish the *fact* of Europe's historical superiority to non-Europe before Jones begins to inquire into the *causes* for this superiority. These assertions are false and occasionally quite bizarre. Jones states flatly that the "energy output" of Europe's people was higher than that of other people in 1500 (p. 3). This means that Europeans in prior epochs were more vigorous and better-nourished. Europe's "real wage" was uniquely high as far back as 1300 (p. 3). In Asia, the standard of living was, by contrast, low. Early Asia, says Jones, *seemed* rich and grand, but this was illusion, based on the grand works of civil engineering and the luxurious living of its ruling classes.

In fact, there is solid evidence that historical Asian civilizations were indeed on a par with or even well ahead of European ones; standards of living were at least as high; and the luxurious life of the ruling classes was an indicator of the society's wealth.[6] Jones, nonetheless, will repeat over and over again the assertion that there was something peculiarly decadent and barbaric about the lives of the rich in Asia. This will be asserted as part of a package of beliefs that can be recognized as the classic myth of "Oriental despotism," which we discussed in Chapter 2. Asian civilizations, in this myth, were a compound of desperate poverty of the masses and disgusting luxury for the rulers. Rule was absolute and despotic. Money was squeezed from the peasants only for public works and ruling-class luxury. The rulers could "squeeze blood out of stones"—the peasants—because "the stones were numerous enough" (p. 5). Europe, by contrast, was always, in one sense or another, democratic. The poor lived better; the ruling class displayed less "splendor" (p. 5). The myth of Oriental despotism is in fact a very old one, a compound of early-modern attitudes toward India and the Ottoman Empire and of self-congratulatory beliefs about the supposedly permanent nature of Europe's democracy. The myth is completely false. Jones merely repeats it in its standard form.

The systematic argument designed to explain Europe's eternal superiority begins with a series of misstatements about the natural environment. Citing old and discredited environmentalistic writings as though they were authoritative, Jones claims that Europe's environment was superior to that of Asia and Africa. Basically this is an argument that midlatitude environments are superior to tropical ones. Says Jones, in hot climates "human energy" is less (p. 6), and there is a truly terrible problem with disease. The old myth that temperature has anything to do with human energy output was really discarded about a half-century ago.[7] The idea that tropical regions are unhealthy is very largely a myth. While it is true that some diseases are more serious when there is no winter season to suppress certain organisms, it is also true that most of the serious diseases have a different epidemiology. Some are more serious in cold weather than in warm. Dry seasons in most tropical regions have much the same suppressive effect as cold seasons elsewhere. Moreover, domesticated animals, along with rats, are the prime sources of infection for many of the most serious human diseases, and contact with these animals is just as intense in colder regions as in warmer ones. And in Jones's assertion that the unhealthiness of the tropics somehow inhibits the development of civilization there is, again, an ancient and discredited myth. Disease did not have that effect.[8] Tropical civilizations flourished during roughly the same epochs and to the same degree as midlatitude civilizations prior to the modern period.

We notice, in all of this, Jones's method of asserting myths as though they were established fact:

> Combined ill-health, heat and malnutrition in the tropics have been shown to cut labor productivity per man by up to eighty-seven per cent, besides raising absentee rates. (p. 7)

None of that has been "shown." Jones's sole authority for this statement is one eccentric and rather laughable essay entitled "The Curse of the Tropics."[9] The entire notion that human bodies (and minds) function less efficiently in tropical environments than in colder ones has long since been rejected.

Many more environmentalistic arguments are put forward in *The European Miracle*. (In *Growth Recurring* some of these arguments are reiterated, and none, as far as I can tell, is repudiated.) Jones admits, as he must, that European soils were not as productive, under rain-fed agriculture, as were the soils of regions of irrigated agriculture in Asia that had adequate water supply. This becomes transformed into a spurious argument for European superiority. Irrigation, Jones asserts, requires a lot of labor, mainly to maintain waterworks. Rain-fed agriculture, though less productive, requires less labor. Therefore European farmers were really more productive, not less so. "The very impracticability of hydraulic agriculture freed a fraction of European energies for other purposes. . . . [Europeans] spent less time on all aspects of farmwork than [Asians] spent on water control work alone" (p. 8). (Again we are given an obscure and valueless source as authority.[10]) The implication that Jones draws is that European agriculture had a greater capacity to produce a surplus beyond the subsistence needs of the farmers, therefore to provide capital for investment in innovations, and so to generate progress in Europe while Asia remained stagnant. If Jones's theory about rain-fed agriculture in comparison to irrigated agriculture were true, European agriculture would have been able to support more people per unit area than irrigated farming systems of Asia; the opposite was the case. And European farmers would have had more free time with which to build social or material structures than Asian farmers; again, the opposite was the case. In fact, there is no evidence whatever, and no logic to the argument, that irrigated agriculture required more labor input in relation to production than did nonirrigated, rain-fed agriculture.

Jones believes that the single most important reason for European superiority, throughout history, has been the marvelous ability that European people have to control their population, as contrasted with the uncontrolled reproductive behavior of Asians. Says Jones: Europeans don't waste whatever progress they attain by simply producing more babies, as Asians do and always have done; instead, they invest the fruits of progress as capital, for future development. The root of this argument is a truly classic assumption that Europeans are more rational than non-Europeans: they are smart enough to practice birth control, whereas non-Europeans either are too stupid to do so or are, like the beasts in the field, subject to uncontrollable sexual urges that override any rational concern to keep their population in check. This argument is softened in *Growth Recurring* but not really discarded: non-Europeans are there declared to be as rational as Europeans in economic matters but not in matters of demography.

In Malthus, and also in Weber, a source of this uniquely European rationality in matters of reproduction was race. Where does it come from for Jones? The story begins, again, with the natural environment. Europe's environment did not permit irrigated agriculture. For this reason, Europeans practiced a very nonintensive kind of agriculture from earliest Neolithic times, thousands of years ago. Their cultural landscape, says Jones, was forest interspersed at wide intervals with fields and pastures. Although, when the main stock of future Europeans migrated into Europe from Asia many thousands of years ago, they tended to live, Asian-style, in communal villages, the new environmental conditions led to a change in settlement patterns. Mainly because farming was extensive, not intensive, populations tended to scatter. Villages broke up first into extended-family household units and then into nuclear-family units, associated with dispersed, isolated holdings. This, says Jones, had two immensely important effects on society, effects which go far toward explaining European superiority thereafter. First, this produced

> the cellular, high-energy, high-consumption life-style and individualist preferences of the Celtic and Germanic tribes. . . . Europeanness lies in the form of the original settlement history, . . . a decentralized, aggressive, part-pastoral offshoot and variant of western Asian agricultural society, moulded by the forest. (p. 13)

Because Europeans (supposedly) lived in rather dispersed settlements, determined by the environmental conditions, they were able to avoid the despotic fate of Asian peasants. Not living in compact masses that a ruler could easily control, they held on to their freedom. So these early Europeans were crude and rude but free and innovative and aggressive—in a sense, embryonic capitalists—and this became one of the roots of "Europeanness."

The errors in this argument are abundant. Most early European settlements were nucleated or linear, in river valleys, along coasts, and so on. Isolated households must have been rare. The pattern of small villages, not subject to kings or large-scale landowners, gave way wherever the Roman Empire held sway. Jones wants us to believe that the northern European tribal roots are more important in European culture and history than the Mediterranean roots, which were based in class differentiation, literacy, Greek and Roman polities, and so on; this argument is not only dubious but, even for the Germanic and Celtic areas, it demands that we consider the basic culture to have been formed before Roman times and early feudal times, and to have been, somehow, different from tribal cultures elsewhere.

Further, the image of massive populations of irrigating farmers in great river valleys is not really correct for most regions of Asia and Africa. Peasants in many regions of Asia practiced rain-fed, nonirrigated agriculture, just as most (not all) European peasants did, and irrigated agriculture was most frequently practiced in small systems and regions, not in huge river valleys on the scale of the Nile and the Tigris–Euphrates.[11] In other words, the model of Asian society that Jones uses is not an accurate one. If there were indeed a causal relationship between upland landscapes with rain-fed agriculture and various intricate cultural effects like individualism, aggressiveness, and family type, then we would find these effects as abundantly in nonirrigating parts of Asia (and Africa) as in Europe. The entire argument is fallacious and unsupported. Apart from a single reference to the virtues of "rain-fed" agriculture, the argument does not reappear in *Growth Recurring*.

A second root, according to Jones in *The European Miracle*, was the nuclear family itself. He asserts that the nuclear family has been characteristic of northern Europe since those days in dimmest Neolithic history when, he claims, the environment required a settlement pattern of dispersed households. He claims next that the nuclear family, as household and social unit, is a basic cultural trait for Europeans. It is more suited to social progress than are extended families. It encourages savings, hence capital accumulation. Most importantly, it encourages birth control, although Jones does not tell us why nuclear families would limit their population and extended families would not. In *The European Miracle* the superiority of the nuclear family as a source of progress is emphasized; it is claimed to be characteristic of Europeans and to be of ancient tribal origin in Europe. It contrasts with the extended (or joint) family, and extended family household, in Asia. "[The] Indian joint family system was a disincentive to save or limit births because the individual could not be sure of keeping any gains to himself or his nearest kin" (p. 193). Actually, it is usually argued these days (and correctly so) that the extended family household is no less suited to savings and ambition than is the nuclear family. Why? All members help one another; they cooperate; loss of a job by one household member is not, therefore, disaster; larger stocks of savings can be accumulated by a larger group, allowing more fruitful investment; ambition embraces the goal of leading not only one's spouse and children to a better life but in fact leading what is in essence a small community, with all the prestige and pride that this implies; and so on. Indeed, among many immigrant groups, such as overseas Chinese, the extended family household often is the core of a successful business. In *Growth Recurring*, Jones accepts part of this argument. But he retains the completely unsupported proposition that the nuclear families, supposedly characteristic of Europe, have a stronger incentive to practice birth control; and this theory he develops into a basic explanation for what he sees as a prime characteristic of "Europeanness": resistance to overpopulation. This Malthusian view is maintained in *Growth Recurring*, as we will see.

Extended family households seem to have been the norm in most of ancient Europe. Early Europeans did not live in isolated households: to do so would have been very dangerous indeed. Farm families usually lived in clustered or linear settlements, and close to one another. Nuclear family households certainly were no more common in Europe, prior to early modern times, than they were outside of Europe. There is very little evidence that nuclear family households were common in northwestern Europe before the early-modern era. Most fundamentally, nuclear family households were not unique to Europe, and they would not in any case have had the effects on history that Jones claims for them.[12]

In making these arguments for the effects of early cultural patterns on later history—unproven patterns, as we have seen, and mythic effects—Jones is aware that some scholars will offer the objection that culture changes, as history progresses, and ancient patterns should not have continued effects in later times. But in *The European Miracle* he insists on the tenacity of long-standing cultural patterns and their permanent effect throughout history. He says, in essence, the culture of ancient northern Europe contained traits and institutions that were uniquely suited to progress, development, superiority. So progress, development, superiority were therefore evidenced by Europeans throughout later history. (In *Growth Recurring*, he completely downplays the significance of culture, except economic culture.) He also expresses the proposition in its contrasting form for non-Europeans: their cultures had traits and institutions in early times that would hold back their development throughout later history. They would breed like animals. "[Population] was permitted to grow without . . . deliberate restraint. Seemingly, copulation was preferred above commodities." They would suffer the yoke of despotism instead of acquiring democracy (thus demonstrating "political infantilism," p. 10). Culture is indeed important and cultures persist, but they do so no more tenaciously in Asia and Africa than in Europe.

Having explored the ancient origins and environmental bases of Europe's superiority, *The European Miracle* next begins a systematic exploration of the forces which sustained and increased that superiority throughout subsequent ages down to the present. These, Jones argues, are:

- continuous technological inventiveness and innovativeness, unique to Europe: not matched by any other civilization;
- a unique urge to expand, to venture, to explore, to discover;
- a uniquely progressive market economy, peculiarly adapted to the rise of capitalism; and
- a unique form of polity and system of states, uniquely suited to continuous progress and to capitalism.

One chapter in *The European Miracle* is devoted to each of these arguments. Let us examine each of them in turn.

TECHNOLOGY

Europeans are technological supermen, according to Jones. Here are some of the ways he expresses this doctrine in *The European Miracle*:

Europe was a mutant civilisation in its uninterrupted amassing of knowledge about technology. (p. 45)

[On] the technological front, the history of Europe [looks] like a persistent drifting advance ... compared with the sluggish nature of other civilisations. (p. 56)

[Europe's was a] traditionally inventive economy. (p. 59)

Ceaseless tinkering is a defining characteristic of [European] culture. (p. 62)

[Given] the inquisitive practicality of many who in other societies seem to have spent their leisure single-mindedly in pursuit of pleasure or at best in impractical philosophising, the persistent advance [of technology in Europe] is not so surprising. (pp. 63–64)

Europe was a peculiarly inventive society. (p. 227)

Europe's technological genius is attributed, as these quotations make evident, to the peculiar rationality, inventiveness, and innovativeness of the European mind. Jones's view in *The European Miracle* is similar to Max Weber's and Lynn White's, although he differs from them in his search for a material environmental cause of the phenomenon. In *Growth Recurring* he deliberately distances himself from Weber's idea that the traits of rationality are somehow peculiarly European. He argues, first, that the notion of rationality should focus on economic rationality, and all people of all societies are rational in their responses to conditions affecting economic behavior; the trouble is, these conditions have always been less favorable elsewhere than in Europe. Second, he argues that the noneconomic cultural factors such as religion and cultural values, which most Eurocentric historians (more or less following Max Weber) deploy as explanations for non-European lack of innovativeness, are really unimportant as factors in economic change or nonchange.

But Jones does not really tell us why Europeans *are* peculiarly inventive, innovative, and so on. In *The European Miracle* he asserts that it has been there "from ancient times" (p. 46), and attaches it to his conception of the freedom-loving, individualistic culture that he claims was the characteristic European culture of the pre-Roman period, rooted in nonirrigated farming. (In *Growth Recurring* he traces it back only to early medieval times, neither affirming nor denying the validity of his previous argument about ancient settlement patterns and the like.) Jones argues that, outside of Europe, farming cultures were burdened by overpopulation (insufficient demographic restraint), by the need to irrigate, and by Oriental despotism, whereas Europeans were free, well-fed, progressive, hence by implication technologically innovative. It must be said again that this is pure mythology. Ancient European tribes were much like ancient Asian and African tribes in all matters here under discussion. All of them were freedom-loving and the rest. Irrigation was practiced only in certain parts of Asia and Africa, and irrigated farming systems did not condemn their practitioners to any of these effects. Indeed, most of the major technological innovations of the period from the late Neolithic down to the Roman era (or rather the Roman–Ghanaian–Mauryan–Han era) came from irrigation-based civilizations: writing, road building, architecture, and much more. Nuclear families were no more characteristic of Europe than of non-Europe in premodern times. Birth control practices of one sort or another seem to have been universal.

During the Roman period in northwestern Europe, peace led to population growth, and "in order to support a growing population . . . the Germanic peoples were given an incentive to turn to technological invention and innovation." The result was a spurt in agricultural technology that gave Europe an immense push forward. We notice first of all that Jones sees population growth as an incentive for innovation in Europe, while in the case of Asian civilizations he considers it to be a *disincentive* to innovation (pp. 158, 169, 201, 215–219). The basic reason, again, is differential rationality. Early Europeans, says Jones,

> were prepared to trade off . . . children for goods. . . . [Unlike Asian males] European males did not practice this immediate division of the spoils of love. By that *restraint* they [held down] population. . . . [In Asia] population was permitted to grow without such deliberate restraint. Seemingly, *copulation was preferred above commodities.* (p. 15, emphasis added)

Asians apparently copulated without restraint, did not understand birth control, so suffered unwanted population growth that (somehow) either prevented technological change or wiped out its fruits. Jones makes much the same argument, though with more restrained language, in *Growth Recurring*:

> Only for Japan and Europe is it usually claimed that peasants controlled family size, implicitly choosing income rather than an additional child. . . . Deliberate population restraint is not reported from the societies of mainland Asia. . . . Perhaps, then, peasant demography was the spring of the trap in which most of the pre-modern world was willingly caught. (p. 127)

Again:

> [Demographic] restraint was . . . conspicuously absent from India and other parts of Asia. (p. 212)

None of this is true.[13] The phrases "usually claimed" and "not reported" simply disguise Jones's ignorance of the pertinent literature.

Jones does not really provide an explanation for this putative difference between sensible Europeans and irrational Asians. The closest he comes is in his quite unsupported and illogical theory about the European family. Supposedly, (northern) Europeans have practiced birth control and have had small families since pre-Roman times. This implied a culture pattern in which economic goals would be set higher than the goals of uninhibitedly increasing the number of children. Supposedly, Asian cultures don't have this pattern. In fact, birth control seems to be a universal in culture, and there is no reason to suppose that Europeans in the past practiced it more successfully than non-Europeans. And the argument that smaller families have higher economic aspirations is vacuous. Jones announces flatly that Europeans, uniquely, have maintained small families since ancient times. "We may indeed suspect" he says, that the pattern goes back to the second millennium B.C. However, the "demographic literature skirts the issue" (pp. 15–16)—here again a device for avoiding the admission that the evidence just is not there.

Jones adds one other putative explanation for the permanent technological genius of European culture. He claims that natural disasters are so common in Asia that peasant families must, somehow, breed more children as a defense against catastrophe. The idea that peasant families in Asia historically had more children in some, presumably unconscious, adaptation to the frequency of natural disasters, responding to "the effects of disaster by maximal breeding" (p. 20), is simply speculation. In any case the claim about natural disasters is not valid. Since Asia's area of settled population is about four times the size of Europe's, we should expect Asia to have four times as many natural disasters as Europe.

So, in the end, Jones does not really develop his argument about the ancient (pre-Roman) origins of what, for him, is a very ancient trait: European technological rationality. From this fact one can infer that Jones finds the roots of this trait, not in the physical environment but in "the quality of Europeanness."

European rationality shares credit with the European environment in Jones's discussion of the technological vitality of the post-Roman Dark Ages and the early Middle Ages. "High, even rainfall and passable summers" gave northwestern Europe a supposed environmental advantage for the production of a variety of food crops along with livestock, the result being "a varied diet" and "[nutritional] advantages over the cereal diets of the older civilisations" (pp. 48–49). To describe this as an environment uniquely favoring "a varied diet," and so on, is just wrong. (It seems to come from Lynn White.) To begin with, Jones has a distorted image of Asian agriculture ("the older civilisations"), which actually yielded as great a variety of foodstuffs as did European agriculture. "The rice landscape," he says, "was extremely undifferentiated" (p. 212), inhibiting regional division of labor and trade. But this is hardly ever the case. Rice dominates certain landscape because its very high food value and high sale value make it the most desirable crop for farmers to grow. Other crops are grown in association with it, in rotation or on adjoining unirrigated land, and livestock feed on rice stubble as well as upland grass. The rice landscapes of historical Asia provided quantitatively high and qualitatively varied nutrition, certainly in no way inferior to the typical European rural landscape. In any case, rice does not dominate the landscape of Asia; no single staple does.

It is a classical prejudice of (northern) Europeans to imagine that their environment is uniquely productive for agriculture. The "high, even rainfall and passable summers," so praised by Jones, very often mean waterlogged soils until very late in spring or early summer, thus a short growing season for many crops. Some regions could not support crop farming prior to the introduction, from South America, of the moisture-loving potato. Total biomass production rates are distinctly limited by the high rainfall, cool summers, and very marked cloudiness of north-western Europe as compared with warmer regions, such as temperate north and central China and north India, and tropical regions, like southern China, most of India, and all of Southeast Asia. The climates that permit highest biomass production, and therefore (in general) highest potential crop yields, are tropical and subtropical, not the midlatitude maritime climates of northwestern Europe.

Jones now begins a simple and traditional account of the supposed cornucopia of technological innovations that spilled out during the Middle Ages and propelled Europe forward in a "persistent drifting advance" ("compared with the sluggish nature of other civilisations"), an advance that supposedly continued without interruption into the modern age. It is important, at this point, that we draw an important distinction between two kinds of accounts of medieval technology. It is perfectly true that some technological advance was taking place, so one can recount that process, add a few romantic adjectives, and so produce what seems to be a description of a genuine technological revolution. (Lynn White and Robert Brenner did precisely that, as we saw in Chapters 3 and 4.) Or one can use different adjectives and make the whole process seem quite slow and unimpressive. Either kind of account makes use of the same facts. So the question is: how to establish a yardstick that will tell us whether technological advance was fast or slow, revolutionary or not. Two such yardsticks are at hand. First, if it can be established that profound cultural change took place as a result of the medieval technological changes, then those changes were indeed revolutionary. Second, if a valid comparison can be made with other societies so that it can validly be said that Europe was moving forward quickly, other societies slowly or not at all, then, again, we have a revolutionary process. Jones tries to do both, and fails.

First, he asserts that medieval technological advance was one of two crucial sources of the economic advance of the Middle Ages (politics was the other). Here he puts forth Lynn White's basic argument, although he qualifies the case to take into account some of the objections that had been made to Lynn White's technologically deterministic theory in the two decades that had passed since the publication of White's *Medieval Technology and Social Change*. Jones sets down White's list of the fruits of the technological cornucopia and adds a few traits of his own.[14] Jones admits that it is difficult to establish whether technological advance was an independent variable (as White insisted) or was, rather, an effect of such things as increased political stability and its economic effects. He concludes that technology was largely but not entirely an independent factor. For our purposes his doubts on this matter are of no great interest because Jones explains political and economic processes—as we will see in a moment—in terms of the same primordial rationality of "Europpeanness" as he does technological processes. In a word: medieval social change is seen by Jones as a complex dynamic process, impelled forward by technology and other factors, all of which are, themselves, symptoms of the basic factor, European rationality.

Second, Jones tries to show that this steady forward progress of European technology was not matched elsewhere, was indeed an element in the medieval "European miracle." In the days when Weber was writing about these matters, this task appeared simple, because Europeans at the beginning of the twentieth century believed that almost nothing of importance was invented on other continents during the Middle Ages. Even at the time White wrote *Medieval Technology and Social Change* (1962), relatively little research had been done on the technological and scientific history of Asia and Africa, and he can be forgiven for the fact that the majority of his claims about European technological inventions in the early Middle Ages have since been shown to be invalid on grounds that the traits in question were invented elsewhere. Jones, for his part, is unaware of the most important research on non-European technological origins, but he cannot completely ignore the existence of such work, so he brings into play a few of his verbal devices.

We are given, to begin with, a generalization to the effect that many of the technological innovations that appeared in Europe in the Dark Age and Middle Ages *of course* were invented elsewhere, and diffused into Europe, but "it was Europe that brought them to a high pitch, employed them productively on a wide scale, and generally in technology and science came to surpass its mentors" (p. 58).[15] This is misleading in several ways. If Europe "came to surpass its mentors" only many centuries later, during the modern era and in particular since the start of the industrial revolution, then the medieval Europeans deserve none of the credit:

there was no medieval "miracle." Jones throws in the obligatory example, China's invention of gunpowder. He believes, as European children are taught to believe in school, that, although the Chinese invented gunpowder, it was actually the Europeans who turned it from a plaything into something serious: explosives for military use. But Needham and others have shown that the Chinese not only invented gunpowder but also used it in cannons at least as early as the Europeans.[16] The first European and Chinese cannons seem to have appeared in the same decade or so, and the possibility that the complex as a whole diffused rapidly from China to Europe cannot be discarded. But Europeans did, indeed, bring the technology "to a high pitch"—*after* the Middle Ages.

Having conceded that some of the novel technological traits were invented outside of Europe, Jones discusses specific traits as though they were indeed invented in Europe, leaving the impression that all of the really important traits were, indeed, invented—or, in any event, perfected—in that continent. Here we have a long list of traits, most of which in fact were *not* invented in Europe or were not introduced only in Europe. The list is long. It begins with the key items from Lynn White's list: leguminous crops, the heavy plow, horseshoes, the horse-collar, and so on. Then come mechanical devices: Jones seems to believe that the whole technology of mechanical engineering is peculiarly European. The windmill and the water mill. (Both were widespread in the Old World and probably were not invented in Europe.) As to clocks, Jones here joins other celebrated Eurocentric historians, among them Carlo Cipolla and David Landes, in claiming, falsely, that clocks were not used at the same time or earlier in other civilizations: Needham, for instance, chronicles very ancient clocks in China.[17] (We discuss this matter in Chapter 9.) Cannon. The printing press. Jones concedes the Chinese origin of printing (actually, it may have been Korean) but claims, falsely, that the earlier Chinese technique "was not as flexible as Gutenberg's. . . . Mass production would have been difficult" (p. 62). This is Jones's springboard for a leap to important conclusions about the effect of printing on "the minds of men"—European men only, not Chinese, who in fact read books, too, in those times.

Still another device is used by Jones to distort the history of technology to make it seem that Europeans were the only miracle workers in the Middle Ages. This is an old one, recently revived by Eurocentric historians. Europeans have known for a long time that early Oriental civilizations had been more advanced than early European ones in many ways, including many areas of technology. The classic judgment was: "Somehow, they attained these heights without the benefit of Christianity, but without God's guidance their civilizations had to stop progressing and in-deed retrogress and decay. With His guidance we came to surpass them." In other words, Oriental achievements were not explained, but it was al-ways insisted that these Orientals had stopped progressing at a certain point in history and thereafter retrogressed. Today, Jones and others em-ploy a modern, secular, and sophisticated form of this old stagnation ar-gument. Wherever technological progress is known to have occurred, and cannot be denied, one quickly adds: *"But it stopped!"* For instance, China developed some sophisticated techniques for textile manufactur-ing in the Middle Ages; this did not lead to an industrial revolution, to a genuine textile manufacturing industry of the sort that developed—five hundred years later—in Europe. But many Eurocentric historians do not just say, "it was a step forward" and leave matters at that. They use the fact as proof that the Chinese did not know how to keep technology moving forward. Most of the examples used by Jones in his arguments about the stagnation of technology outside of Europe are basically of this type. They ask why medieval Asia did not continue to progress to an in-dustrial revolution, ignoring the fact that the real industrial revolution took place centuries later, and took place after most of Asia had been crushed by European conquest or subjugation under conditions of mod-ern colonialism.

Jones's discussion of technology, overall, is designed to show that Eu-ropeans, and no others, made technological advances in the Middle Ages, and to put this forward as an important part of the foundation for his theory of the "European miracle." He fails to make his case.

EXPANSIONISM

Europeans, according to Jones, have what amounts to a natural tendency to enlarge the frontiers of their society by exploration, discovery, con-quest. The Age of Discovery, the age of Columbus and Vasco da Gama, was merely one stage in

> an old endeavor to pierce the void. . . . European society had been pulsating and probing at its bounds for a long time before that, at least since the tenth century . . . or earlier still if the Viking crossing of the North Atlantic be in-cluded. (p. 70)

This view, it should be noted, is commonly encountered in Eurocentric historical writing. Usually it is seen as one dimension of European "rationality": Europeans have the urge to invent, to innovate, to progress, and so to inquire, explore, discover, and so on. Jones is not the first to claim that the expansion of Europe to America, Africa, and Asia was a reflection not of conditions that prevailed across the Eastern Hemisphere in the fifteenth and sixteenth centuries, but of an old tendency within "vital, expansionist European culture," (p. 75) because Europe was just (in Jones's words) that "kind of society" (p. 71). Jones is quite conventional, also, in referring to the Vikings, the Crusades, the expansion eastward toward Siberia, the reconquest of Iberia, and so on, as moments in this "continuum of expansion" (p. 75).

Jones does not try to explain Europe's urge to expand. I infer that he sees it in the same light as Europe's overall habit of inventing and innovating. He begins his discussion of the expansion with the tenth century, associating the beginning of the Crusades, and the like, with the establishment of peace in northwestern Europe after the Dark Ages. The expansionism itself emerged in the High Middle Ages, and we are left to infer that it was latent before that time.

Jones concedes that the "Islamic world" (undifferentiated) was also expansionist, conflating in the process the Turkish conquest of Egypt, the Mughal entry into India, and the "Moorish" conquest of Spain. But he dismisses this as unimportant on the grounds that, first, Islam expanded in areas with lesser resource endowment than the New World (in this he is of course right); second, Islam was not the kind of society that could make use of the fruits of conquest (wrong); and third, Islamic society in any case declined—the standard refrain: "but it stopped." Jones utterly ignores the fact that Chinese society and, indeed, many other non-European societies also expanded when internal and external conditions were favorable: political power, economic development, the character of the societies at their frontiers, and so on. Chinese expansion is discussed as though it were a sluggish peasant migration. The great voyages of discovery led by Admiral Zheng He (Cheng Ho) in the period 1417–1433 are mentioned in a different context; what interests Jones about them is what he perceives to be evidence of Chinese stagnation and non-progressiveness in the fact that the voyages stopped happening: "they stopped." So, in all, Jones leaves us with an image of Europe as a special kind of society, given to the miracle of permanent discovery and expansion.

If we do a proper comparison of European and other Eastern Hemisphere societies in the Middle Ages, it becomes apparent that all of the powerful and advancing ones were also expansionist where that was feasible. The Crusades reflected no ineffable expansionist tendency: they had definite concrete objectives. The Vikings were much like other roving seafarers, such as the Polynesians. The medieval discovery of some of the Atlantic islands (Azores, Canaries, and so on) had contemporary parallels in the Indian Ocean and the Pacific Ocean. And so on. The argument that there was something unique (and miraculous) about the European expansion just doesn't hold together.

THE FREE MARKET ECONOMY

Since capitalism arose in Europe, and since capitalism, once developed, possesses a market economy—a free one in the sense that transactions are settled and workers are free to move about without much interference from political authority—it would be quite uncontroversial to argue that the development of a free market economy was taking place during the Middle Ages in Europe; part of the overall growth of the commercial economy, and the gradual rise of capitalism out of feudalism, had to be the evolution of a market system, a market economy. Jones devotes one chapter in *The European Miracle* to the growth of the free market; his purpose is to show that this growth was both unique and (figuratively) miraculous.

The root of his argument is the conception of ancient European society as one imbued with "individualism" (the "individualist preferences of the Celtic and Germanic tribes . . . decentralized, aggressive" [p. 13]) and with rationality. Throughout *The European Miracle* we are repeatedly told that Europeans traditionally and since ancient times have possessed special qualities of individualism, innovativeness, initiative; that they are in a sense natural capitalists. Jones does not actually say that non-Europeans are not natural capitalists, do not employ rational economic calculation; rather, one gains the impression that he feels that an urge toward capitalism was in the veins of all human beings everywhere but Europeans were more precocious than everyone else. This is spelled out to some degree in *Growth Recurring*, where he argues that economic rationality is a cultural universal, but environment and the state conspire to prevent rational individuals from getting ahead in non-European places (more on this later).

Jones concedes, as he must, that other cultures also possessed extensive trade in early times, but he asserts that their sort of trade was not the sort that leads toward capitalism. In Jones's view, European trade was truly "free" from the early Middle Ages onward. Asian trade and markets were, by contrast, under political control during all periods. In the chap-

ters devoted to Asia (to be discussed below), he supplies anecdotes supposedly showing that Asian despots rigidly controlled, manipulated, and
bled the trade that occurred in and among their empires. But all of this is
historically untrue. In all medieval societies there was more or less involvement of the lords and kings in commerce, but it was certainly no
more pronounced in Asia and Africa than in Europe. Jones is well aware
of the true degree to which European merchants were able to pursue their
commerce in spite of political interference, both within and across the
boundaries of polities. He seems unaware that Asian and African merchants did the same (and I find no real improvement in *Growth Recurring*). The evidence on this matter is very strong indeed.[18] (We may
recall the famous comment by Tomé Pires, the Portuguese chronicler,
soon after contact was made with Indian merchants: "They are men who
understand merchandise; they are . . . properly steeped in the sound and
harmony of it," and "those of our people who want to be clerks and factors [that is, traders] ought to go there and learn, because the business of
trade is a science."[19])

Jones also repeats the traditional falsehood that Asian trade was essentially limited to low-bulk, high-value items, basically luxury items,
whereas European trade consisted mainly of bulky and utilitarian commodities. He draws the implication that these sorts of commodities were
closer to the real needs of a developing economy than were frivolous luxury goods, so that European trade was pregnant with capitalism whereas
Asian trade was not. (This is qualified slightly in *Growth Recurring*: Song
China had such trade, but at no time prior to modern times did international trade in Asia—he thinks of it as inter-empire trade—have the progressive qualities found in Europe's international trade.) He also suggests
that the luxury trade of Asia was somehow naturally associated with the
decadence of life of the Asian ruling classes and their natural unwillingness to countenance economic development. This is again false. Massive
bulk trade was characteristic of Asia no less than Europe. Rice went from
India to Iraq and from south China to north China. Lumber went from
Burma to Malaya. Iron from East Africa to India. And so on.[20]

Environmentalism is again introduced at this point, to reinforce
Jones's theory about the uniqueness of Europe's "free" market economy
from early times. "The peculiarities of European trade arose because of
the opportunities of the environment" (p. 90). Europe's environment,
says Jones, was uniquely varied in climate, geology, and soils, providing
natural complementarities of resources for trade. Also, transport costs
were low since Europe possessed a "long, indented coastline relative to its
area and . . . good navigable rivers" (p. 90). Back in the days when Euro-

pean geography was dominated by environmental determinism, this argument was a very popular explanation for the greatness of Europe, but it can no longer be accepted. In fact, the variety of environments in Europe is no greater than that in China, India, Africa, and the Middle East. In China the north–south cline in temperature correlates with a change in crop types at least as great as that in Europe (which in fact has neither true tropical conditions, like Hainan, nor true desert conditions, like Xinjiang, nor true high-altitude agriculture like Yunnan and Tibet). Both in China and in India there is a sharp differentiation between wheat regions (cooler, drier) and rice regions; in Europe, by contrast, wheat was a staple wherever it could be grown, north, south, east, and west. And so on. So much for the false argument about Europe's unique environmental variation and the supposed stimulus it gave to European trade.

As to the argument about peninsulas and rivers (which geographers used to call the "capes-and-bays theory"): most commodity movement in Europe (except in the Mediterranean region) seems to have been by land, not by sea, as was the case also in Asia (except on the immensely long coastlines of India, mainland Southeast Asia, and China, and the archipelagos of insular Southeast Asia). And the rivers of Europe are neither more nor less useful for commerce than are the rivers of Asia: the Ganges, the Mekong, the Yangze, and others. So, again, the argument from environmental determinism in support of the "European miracle" thesis does not hold weight. In *Growth Recurring* Jones says little on this subject, but he seems not to have changed his environmentalistic views.

I believe that Europe's market economy evolved in a somewhat similar way, and at roughly the same tempo, as the market economies of Asia and Africa. The divergence began after 1492, when the conquest of America brought riches to the incipient capitalist sectors in Europe, giving them hegemony and the power to begin the process of defeating competing merchant communities in Africa and Asia. Europe's medieval economy was in no sense unique, much less miraculous.

STATES AND NATIONS

The political development of Europe in medieval and early modern times is fairly well understood and, in its general outlines, fairly uncontroversial. Jones, however, twists the facts around to make it seem that this political development was miraculous—indeed, that it was near "the heart of the European miracle" (p. 124). He claims that the modern European state and the modern system of states was, in essence, clearly present in the medieval period. But in fact the medieval political landscape was chaotic: a mosaic of several hundred partially statelike political entities, and nothing that could conceivably be called a system of states existed.

There is wide agreement, though not consensus, about the basic process that led to the crystallization of the system of states and the formation of modern nation-states. It ties the process to the rise of capitalism. Most historians argue that economic development out of feudalism was facilitated by a political infrastructure with certain properties. It was important, though not really necessary, that there be a rather large space within which economic processes could circulate without serious barriers, a "national economy," consisting of a single zone of labor circulation, a fairly large market for commodities, and so on. It was important that the state surrounding this economy be powerful enough to protect the economic interests of citizens. It became important, rather late in the process, to weld the citizenry together into, if not a common culture, at least a consensual community, with inculcated values favoring the economic development of capitalism; thus, into a national state or nation-state.[21] The system of states arose partly because of the need to define the rules of intercourse among these newly integrated polities, thus keep the peace, and partly to establish rules for international movement of people and commodities. But all of this took place long after 1500—after the end of the Middle Ages.

Jones disagrees with this. He argues that the European system of states was, in essence, foreordained by the physical environment of Europe, that it came into being early in the Middle Ages, and that it was peculiarly and quintessentially *European*: that is, both the nature of the state and the form of interaction among states was something that could not have arisen in any other civilization and continent. It was near to the "heart" of the European miracle, not a product of other historical forces and definitely not a late development reflecting the rise of the capitalist economy.

His argument runs about as follows. The environment of Europe is naturally divided into ecological core areas, generally small zones of highly fertile soil in which dense populations developed in prehistoric or pre-Roman times. Thus far the argument is a conventional one among geographers, archaeologists, and historians.[22] Jones now departs from convention in three big leaps in his argument. First, these ancient core areas were really the embryonic states of later times: later evolution is somehow secondary, or perhaps teleologically foredetermined in the ancient cores. Second, he jumps, illogically, from the fact that there were very many such core areas to the proposition that there were a few, some-how natural, cores of the modern European states, so that he can argue that the pattern, location, and even boundaries of modern states some-how have a permanent definition in the natural environment. (He builds a romantic picture of states, as they grow, expanding outward into the forests, swamps, wasteland around them until they meet neighboring states that are doing the same.) Third, we have a truly magical transfor-mation: the core areas somehow grow at about the same pace so that there emerges what Jones describes as a "grid" of states. This, in turn, ex-plains what Jones considers to be three absolutely fundamental features of the evolving European political system: the fact that strong states arose; the fact that a multiplicity of states remained, instead of a single European state comparable to the empires of (for instance) China; and the (putative) fact that the European states formed themselves into a genuine interstate system, a "states system," very early and permanently.

> There appears no *a priori* reason why a states system alone should have in-troduced the world to sustained economic development. . . . [We] need first to account for the existence of such a system in Europe, and Europe alone. . . . It seems to have been based on a characteristic of the environ-ment. This . . . was the scatter of regions of high arable potential set in a continent of wastes and forests. These regions were the "core-areas." (p. 105)

> Enough states were constructed each about its core and all of a similar enough strength to resist . . . conquest and amalgamation: a single unified European state. . . . There [were] a large enough number of approximately similar states to preserve the shifting coalitions that successfully opposed control by a single power. (p. 107)

All of this is designed to lay the groundwork for a theory that will explain how politics played a central role in the economic miracle of Eu-rope's development. The theory, which Jones elaborates in *The European Miracle* and restates in *Growth Recurring*, is really quite simple. Europe would not have developed as it did if Europe had been politically unified under a single empire, like the Chinese empire. On the other hand, there had to have been certain kinds of unity across the continent permitting development and transmitting its effects throughout Europe. Jones argues that what he calls "the European states system" emerged, quite early, and then remained in place, quite firmly, as a political mechanism allowing Europe to have all of the advantages of unity with none of the disadvan-tages (as Jones perceives them to be) of an imperial political system.

This theory is actually a rather traditional one, drawing from the old idea of "Oriental despotism," the idea that Asian civilizations were held in thrall by despotic, centralized imperial polities that kept the citizenry in poverty and which, with capricious and irrational decisions emanating from a corrupt and avaricious and decadent bureaucracy and ruling class, prevented any forward progress toward modernity and economic development. Obviously, says Jones, Europe would not have developed as it did if an imperial state had been in place, snuffing out all sprouts of growth. This kind of reasoning would be utterly uninteresting if it merely argued, "The only kind of political system that would have permitted development was the political system that *did* permit development"—the argument that whatever happened had to happen, and had to happen in precisely the way it *did* happen. Actually this is the kernel of Jones's argument, although there are embellishments. These mainly surround two concepts: the wonderful thing called the "states system" and the awful thing called "empire."

The essential elements in the European miracle, according to Jones, were, first, reproductive restraint; second, technological innovativeness and other rational and progressive mental attitudes possessed uniquely by Europeans; third, a uniquely favored endowment of natural resources; and fourth (in consequence), a social environment that encouraged or at least did not interfere with the European's natural tendency and desire to progress, modernize, develop, and so on using these mental and natural endowments. His theory of the "European states-system" concerns this last.

Jones states, at this point, that he agrees with those who say that the best social environment for progress is one in which there is little restraint on economic activity, that is, a laissez-faire environment. That, he says, was the nature of the medieval economy of Europe: it was comparatively free. And this freedom for incipient capitalism was somehow also freedom in general, that is, incipient political democracy.

Next, Jones says, in essence, *obviously* empires would not permit capitalism to have this kind of (necessary) political freedom; therefore an imperial state would have prevented economic development; therefore all of the non-European civilizations that might otherwise have developed could not do so because all of them were empires. (This is qualified in *Growth Recurring*: not "all," but all except, briefly, Song China and, later, Tokugawa Japan.) His support for this assertion is an attempt to supply evidence that empires did not permit economic development. But instead of evidence we are given a dreary list of all of the prejudices that Europeans have held about non-European political systems of the ancient, medieval, and modern period. Here is what he has to say about the nature of imperial government in general:

> Imperial politics were typically unstable. Unchecked, unresponsive, unrepresentative influence persisted in the hands of those who had the care of the young emperor, often a class of eunuchs. The palace atmosphere was too often a stench of vice, treachery and triviality. It is easy . . . to impute purpose to what was frequently the sway of spoilt and vicious children imbued with total power. [David] Landes remarks of Muslim history, "the male rulers read like an oriental version of the Merovingian snakepit." Emperors were surrounded by sycophants. They possessed multiple wives, concubines and harems of young women, a phenomenon that may have been less the perquisite of wealth and power than the assertion of dominance relationships, the propensity to use people as objects. The amassing of households full of slaves for display purposes rather than work may have had a similar ethological significance. Great attention was paid to submission symbols, kneeling, prostration, the kotow, in recognition of the emperor's personal dominance. (p. 109)

Jones then adds, characteristically, that there are some "counterparts" of this sort of thing in Europe, but in "reading the literature," he gains the "opinion," which can be "quantified," that

> excessive consumption and debauchery and terror were much more prevalent in the empires of Asia and the Ancient World than in the states of Europe. (p. 110)

There is more than a hint in all of this that Jones considers Oriental rulers to have been basically irrational (or mad). In the chapters of *The European Miracle* that deal with Asia per se—we will discuss them in a moment—we are treated to more comments of this general type, spoutings of classical European misconceptions about Asia put forward as though they were authoritative statements grounded in scientific observation.

Asian imperial ruling classes were not much different, in matters of debauchery and the rest, from European ruling classes of any given epoch, allowing for the fact that many Asian civilizations were larger in size and economically richer than European ones during the Middle Ages and therefore their rulers tended to be more lavish in their lifestyles, and freer to indulge in caprice, than were European ruling classes. It is a simple enough matter to assemble a number of Asian traits that traditionally have seemed rather nasty to Europeans (sometimes in fact *are* nasty), then to assert falsely that these traits are characteristic of Asia, and then to claim that the picture thus painted is in fact the real picture of Asia throughout its history. This is an old device, and it is used by Jones in a very traditional way. And it is really the essential grounding for Jones's theory about the political reasons for the European miracle. If Europe had had an *empire*, he says, Europe would have had all of this nastiness. And it would have hobbled economic development. How so? Jones paints a picture for us of what is really a fantasy: that despotic Asian governments hobbled the activities of merchants and entrepreneurs:

> [Those] who sought to do business in these regimes did so on sufferance, un-protected by law, and at their daily peril. (p. 122)

It is not correct to argue, as Jones does here, that the imperial govern-ments of Asia—and let us not forget Africa, although Jones does so—were more hostile to private economic activity than were the contempo-rary kingdoms of Europe. This is an old error that modern scholarship has fairly thoroughly pushed aside.[23] Until the early-modern period, the ac-tivities of protocapitalist communities in countries like India and China were neither more nor less hindered by political obstacles than were those of their counterparts in European kingdoms. And in fact Jones can offer no concrete examples to the contrary. The closest he comes to an example is a casual reference to a decision by the Ming court in 1480 not to renew the long-distance voyages that had been undertaken a half-cen-tury before under Admiral Zheng He. But these voyages were state enter-prises, not private ones. Private trade, domestic and foreign, was going on at a brisk pace under the Ming emperors, whose sporadic efforts to con-trol international trade did not squelch the vigorous private enterprise and economic development that was under way in China during that pe-riod.

At no point in his discussion of the contrast between Asian empires and the "European states system" does Jones mention the fact that many non-European societies were governed by non-imperial governments, usually kingdoms rather like the European ones. There were kingdoms, large and small, city-states with various legal governance forms but con-cretely under the control of merchant communities, even a few repub-lics.[24] If Jones's argument that empires are inimical to economic develop-ment were valid—though it is not—we would have to ask: why would the nonimperial states of Asia and Africa not have the virtues of the states of Europe?

We come then to the European "states system." Says Jones: Europe did not acquire the burden of a single empire for three reasons, two ex-plicit, one implicit. The implicit reason, already discussed, is Asian demographic irrationality ("the links between fly-trap economics"—governmental rapacity—"and population growth"[25]). One explicit reason is the physical environment of Europe, which, according to Jones, is divided into natural core areas surrounded by wasteland, core areas that naturally form the centers of future states. The landscape picture painted here is of a continent with very great barriers separating its various "core areas," and we should note before we proceed that this is really a fantasy. Indeed, the Alps and Pyrenees are such a barrier between the Mediterranean region and the rest of Europe. But there are no barriers of any consequence—taking into account the military and transport technologies of each era in history—across central and northern Europe, from France to Germany to Russia, and down through the Danube Basin into the Balkans. Or, to be more precise: there are low mountain ranges and occasional swamps, but in the context of Eurasia these are truly minor topographic features. In much of Asia the barriers are much greater, and there indeed the effects of such limits to accessibility have in some cases been very great. For instance, the development of distinct states in Burma, Thailand, Cambodia, and Java had something to do with mountain systems, insularity, and the like. If it were true that environmental factors of this sort produced Europe's "states system," there would have been "states systems" in many other places, notably Southeast Asia, and China would have had, if not a "system" of states, at least two distinct states, one in the north, the other in the south.

The second explicit reason why empires flourished in Asia is developed in *Growth Recurring*. Much weight is placed here by Jones on the Mongol invasions as a cause of Asia's nasty empires. Supposedly, the fact that at one period in history most of the states of mainland Eurasia were conquest states, governed by an alien ruling elite, resulted, teleologically, in the permanent condition of Oriental despotism: all future kings and empires governed in the way conquerors did before them, paying little or no attention to the needs of the citizens. Thus "for systematic reasons, institutions in the conquered societies may have behaved defensively, become conservative, and reduced the chances of recovery."[26] There are two straightforward answers. First, for Jones as for most other Eurocentric historians, Oriental despotism is supposed to go back much further in history than the Mongol period. Second, it is hard to give credence to a theory that claims that a relatively short-term regime will have political and cultural effects that last for many hundreds of years.

The European "states system," says Jones, was itself a "miracle" (pp. 104, 108, 124, 125). Jones marvels that the "grid" of independent political entities remained in place, stably, from ancient times on down to modern times without collapsing into a single imperial state or fracturing into many tiny polities. Actually, there was no real system of states before the late Middle Ages, and the real "system," as well as the real stability of most of the individual states, emerged in the modern era, well after 1492—and, be it noted, well after the period when the events described by Jones took place, the period of a developing internal market, burgeoning trade, and the growth of towns and a protocapitalist class. Jones's argument, one should recall, is that the states system provided the environment that permitted all of these things to happen. The cart therefore precedes the horse.

It remains only to mention one or two subordinate myths that Jones introduces into his theory of the miraculous European "states system." He imbues the system with (figuratively, of course) miraculous powers. In an argument that defies logic, he claims that there will be more diffusion of ideas and migration of skilled workers in such a system of separate states than in a single empire within which there are no political boundaries. According to Jones, European states in the Middle Ages were more democratic and less despotic than Asian ones, and so did not persecute travelers or hinder the free flow of the ideas, which were essential to technological and economic progress. But travel in Europe was in fact severely constrained by political barriers, which generally did not exist within large imperial polities, and the diffusion of ideas was certainly hindered in the same ways as well as by language barriers. And European states were not more democratic than Asian states: Jones merely asserts these traditional ideas without evidence, and indeed evidence goes against this theory, not with it.

Second, Jones claims that the development of the European states system was facilitated by a common culture, and he insists that Europe at all stages must be viewed as a cultural whole. Many other writers have made this point, usually pointing to the unifying force of Christianity and the church structure as the basic cause or causes, with attention also to the legacy of Roman rule over much of the continent. This is not in dispute. What is in dispute is Jones's claim that comparable unity was not to be found in many other parts of the world (he concedes the point only for China). Jones wants to show that cultural unity was the counterpoint to political plurality (the "grid") and that this produced, somehow, the kind of civilization that had all the advantages of unity and all the advantages of diversity with none of the disadvantages of each. This is plainly wrong.

When we examine Jones's theory about Asian empires, then, we find that it has nothing to do with size and everything to do with a conception about the *nature* of politics in non-European societies. This is nothing more than the old and discredited idea of "Oriental despotism." Asians, and indeed all non-Europeans, *naturally* suffer nasty, despotic, capricious, irresponsible, evil governments. Only Europeans understand and thus enjoy freedom. Why? Because they are Europeans.

PRIMITIVE AFRICA

Jones turns to the world "beyond Europe" (the title of Chapter 8 of *The European Miracle*) and comments piously:

> Comparisons, or contrasts, with other civilisations are essential for an assessment of Europe's progress. Otherwise conjectures based on a winnowing of the European historical literature are uncontrolled. . . . The comparative method offers the [best] hope for a test of significance. (p. 153)

But what Jones has to offer is hardly an example of the comparative method, much less a test of significance. It is mainly a long and dreary sequence of negative statements, most of them fallacious, about the societies of Africa and Asia in times past. His "comparison," then, is an attempt to show that non-Europe did not have the potential for development prior to its colonization by Europeans; that non-Europe actually was moving backward, not forward, at the time colonialism began. In the following paragraphs I will list a sample of assertions about Africa and Asia that one finds in *The European Miracle* and will show them to be false.

Africa is given the briefest possible treatment: a four-page recitation of colonial-era myths about the continent and its people, designed to show that Africa was much too primitive to have any potential for development. Jones begins with the myth that Africans are close to nature:

> In Africa man adapted himself to nature. The hunter felt part of the ecosystem, not outside of it looking in with wonder, and definitely not above it and superior. After all, there were large carnivores who sought man as prey. The most evocative symbol of this ecological oneness may be the honeyguides (*Indicator* spp.), birds commensal with man. They fly, chattering loudly, ahead of bands of hunters, leading them a quarter of a mile or more to the tree hives of wild bees and feeding on the wax after the men have broken open the hives and taken the honey. (p. 154)

The image is of a continent of primitive hunters, with a hint that they are closer to the animals than are other groups of humans (not "superior").[27] Actually, few Africans are hunters: this is a classical stereotype.

Jones then lists the signs of primitivity. Africans did not have knowledge of the wheel. (Completely untrue.) They didn't have the plow. (Untrue for some areas; true elsewhere because alternative farming techniques were more productive.[28]) Africa "had no major direct influence on the other continents, except maybe as a source of slaves" (p. 153). (Africans domesticated many of the important crop plants, and may have been independent inventors of ironworking and steel-making. Early civilization in Upper Egypt was African. Africans traded as equals across the Sahara and the Indian Ocean. Slavery was *not* distinctively African. Africa's influence on other regions was profound and constant.) Jones now damns with faint praise: "Certainly all was not barbarism. There were towns of some size in West Africa. . . . From time to time large states did emerge" (p. 154). (There were, in fact, great cities and great states in medieval Africa.) And overall, says Jones, Africa had "no stable combination of powers that could erect a common front against Arab or European slavers" (p. 154), meaning that Africans were too primitive to have state power sufficient to resist slavery. (This is true only for the period of the modern slave trade and colonialism, the seventeenth to nineteenth centuries, when many non-European societies, in Africa, Asia, and America, were unable to match the power of an already developed and powerful Europe. The slave trade generally did not reach the great inland civilizations.)

Why was Africa so backward and primitive? Says Jones, incorrectly: Africanists who address this problem blame it on the natural environment. Some say the environment was too lush, Others say it was too harsh. Jones thinks it was both: too dry in some areas, too humid and tropical in others. In the drier areas, agriculture was not productive. (Not true.[29]) In the wetter areas "living was easy," but "there is always a dry season." (A dry season is a help, not a hindrance, in most parts of the humid tropics.) Shifting agriculture, he says, was the African adjustment to these soils, and shifting agriculture destroyed the environment ("the land was not given enough time to rest," p. 154) and was not productive. None of this is true.[30]

Next comes a brief, distorted history of settlement and population. "Negroid" people in small numbers had spread out over most of Africa in historic times, and had not reached South Africa when Europeans settled it. (Not true. Africans were there first.[31]) The population of Africa, states

Jones with neither evidence nor authority for the statement, was small and was insignificant in comparison with other continents in historic times. (Not true.[32]) The resources were poor. (Not true.) Transportation was difficult. (Not true.) African chiefs squandered their profits on "luxury items," and so trade could not transform society. (Nonsense.)

> At the root of all this seems to have been the infertility of soil; pervasive insecurity as a result of conflict and slave-raiding ... and a hot environment. ... The defects of the environment [struck] so close to the heart of economic life that *it is not clear what indigenous developments were possible.* All told, there was no development of the African economy to set alongside that of Europe in the Middle Ages and after. (p. 156, emphasis added)

None of this can be taken seriously. Nearly all of it has been disproved by scholarship or has been shown to be unsupported myths from colonial times, and the rest is idle speculation. Indeed, Jones cites no authority for these statements. I believe that, in the Middle Ages, historical progress was as intense and fruitful in Africa as in other continents.

> The *very* long-term economic history of the world was thus acted out in Eurasia. (p. 157)

That is: Africans, Native Americans, and Oceanians played no important role in history.

BARBAROUS ASIA

Why did Asia not develop as Europe did? Jones takes great pains in *The European Miracle* to explain why there was not the remotest possibility that Asian civilizations would develop and modernize. (As we will see, he retreats just a bit from this position in *Growth Recurring*.) The explanation is very elaborate: it takes up about one-fourth of the pages in *The European Miracle*. But it is entirely unoriginal, entirely in the mold of classical colonial-era views about the Orientals, so we do not have to scrutinize the arguments for Asia's inferiority in the same systematic way that we did in assessing the arguments for Europe's superiority. Instead, I will look at each of several *kinds* of explanation, giving examples as we proceed.

The inferiority of Asia is explained by Jones in terms of two fundamental kinds, or categories, of deficiencies: (1) a psychological deficiency, consisting of irrationality in matters of intellectual vitality and innovativeness, combined with a sort of moral failing in attitudes relating to the desire for progress, resistance to domination, will to forgo animal pleasures, and the like; and (2) an inferior natural environment, or more precisely an environment not conducive to economic progress and growth. The effects of these failings (and some lesser ones), and therefore the effective reasons for Asia's nondevelopment and nonprogress, are (1) uncontrolled population growth and (2) bad government. I will now summarize each of these arguments in turn.

We begin with Jones's assertions about the Asian mind, assertions that seem quite peculiar yet are, for the most part, taken from the glossary of traditional colonial-era European ideas about Asians. None of these assertions is supported with real evidence or scholarly authority, nor could they be. They are indeed so bizarre that no comment is needed. A mere listing will suffice.

Orientals do not think logically. There is "relative absence of the empirical enquiry and criticism of the Graeco–Judeo–Christian tradition" (p. 161) and "lack of a crisp tradition of logical debate" (which may explain the "failure" of Asian science). "The notion of a consensus in interpreting nature may have seemed absurd" (p. 162)—that is, Asians may not even have had a concept of scientific verification. They tended to be uncreative: "[Despotic] Asian institutions suppressed creativity or diverted it into producing voluptuous luxuries" (p. 231).[33]

Orientals have (or had) various attitudes and values that clearly inhibit progress. "Oriental philosophies [emphasize] emotions, values and cosmologies" at the expense of empirical thought (p. 161). Orientals are lazy. They have a "love of luxury" (p. 170), and like to purchase frivolous luxury items, aphrodisiacs, opium, "kingfisher feathers . . . precious stones . . . drugs no modern pharmacopoeia would own" (p. 164). They have a "servile spirit" (this an approving quotation from Montesquieu); their armies lack "tough" petty officers (p. 167); they are submissive, passive, and "inherently unresistant to autocracy" (explicitly referring to Muslims, implicitly to Asians in general) (pp. 182, 176). They are, as a rule, introverted, inward-looking; they are "increasingly immobile societies undergoing 'curious experiences'" (p. 170), given to a self-imposed "isolation" (p. 170), and lacking an urge to explore (pp. 168, 177, 203, 231). They are given to senseless warfare (pp. 169, 196, 201), do not have a written legal system (pp. 164, 188, 197), do not have a concept of political boundary (pp. 167, 194). There is much thievery and piracy (pp. 189, 199, 209, 229–230).

Islamic society was for a time innovative, borrowing technology from other societies, but this came to a stop. The Ottoman Empire stamped out original thought. It produced unreason, intellectual backwardness and retrogression, a "mist of obscurantist thought" (p. 183). Ottomans didn't even know "the elementary facts of geography" (p. 184) and couldn't make decent maps (p. 179).[34] Ottoman rulers were "degenerates," "drunkards," "mental defectives," "lechers," ruling with despotism and terror (p. 186–187), their "philosophy" being theft and despoilment, against which there was "no legal shield" (p. 187–189).

Indian society was socially and psychologically "frozen" (p. 192), with values that were deleterious to economic progress. Religion was invoked to sanction all acts, but the advice of religious counselors was "malicious or random" (p. 195). The Mughal rulers (like the Ottomans) were degenerates, running society for their own benefit, given to "voluptuous selfishness" (p. 196), harems, jewels, menageries, intrigues, and treason. The state was purely predatory. Technology was "almost stagnant," not even copying from abroad (p. 199). Here, again, there was no law: "No written legal code existed" (p. 197). (This last statement is not only incorrect—Indian written law goes back thousands of years—but, given that Jones is a historian, remarkable for the ignorance it displays of historical fact.) Demographic behavior was irrational: "[a] similar calculus . . . underlay human demographic strategy and veneration of the cow" (p. 19).

China was technologically somewhat inventive and innovative until the Middle Ages, when progress stopped. There was thereafter a "retreat" (p. 203); mechanical contrivances were dismantled; some skills were even forgotten. Chinese became "inward-looking" (pp. 203, 216, 220). China "backed away" from technology, from trade, from exploration (p. 203). Technological development stopped even in agriculture, and only the irrational cutting of irreplaceable forests, the fortunate arrival of New World crops like maize and potatoes, and the cultivation of new land saved the Chinese temporarily from disaster. (But for Europeans, the cultivation of new land was progressive: "the availability of extra-European territory provided an essential safety-valve," p. 108). Cutting down the forests was "one of mankind's greatest acts of ecological stupidity," stupidity that led to "soil erosion, gullying, silting and floods" (p. 213). Peasants were given to "envy and suspicion" (p. 206) and were stupid as farmers (pp. 212–217), stupid also in preferring "maximal reproduction" over "affluence" (p. 218). The state was "despotic" (pp. 159–166, 206, 210–211, 221–222, 231), a "revenue-pump" for the rulers, providing no services (p. 206). There was a love of luxury, an attitude of "empty cultural superiority" (p. 205), a corrupt, venal, parasitic ruling class, given to displays of ethological dominance (pp. 209–210) and to murder and torture (p. 207). Chinese had "anti-social customs" (p. 7) and were diseased.

Environmental factors are also invoked by Jones in his efforts to explain Asia's (putative) lack of ability to progress down through history. Some of his assertions on this matter were discussed previously: his belief that humid tropical environments are nasty, his false notions about the need for irrigation in Asian landscapes, his mistaken idea that Asia is peculiarly afflicted with natural disasters, and the like. We now add some additional assertions, none of them correct.

Asia, Jones says, is mostly made up of large natural lowlands, and this tends to favor large imperial states like China. (China is as dissected topographically as Europe.) In Southeast Asia, core areas are fertile but are separated from one another, and this produces "political weakness" (p. 166). Jones does not notice that this thesis exactly contradicts his theory about European core areas and the wonderful grid of medium-sized states that evolved from them. Further contradiction appears when Jones asserts that India, one of his "empires," is fractured into isolated regions "separated by wide belts of deserts, hills or jungle," as a result of which the "country seemed to fly asunder at the touch" (p. 194). (How then can it be described as an imperial state? You can't have it both ways. In any case, the geography here is false: India is not fractured in this way.) Asia has poor fisheries resources. (Not true.) China's abundant internal reserves of land somehow favored extensive development and discouraged technological intensification. (Not true.) This internal frontier also explains why China "could survive intact and at the same time remain inward-looking" (p. 220). (No logic here.) It should be said, however, that Jones's environmental determinism focuses less on Asia's putative deficiencies than on Europe's supposedly marvelous natural endowments, which we have discussed already at sufficient length.

Nor do we need to say much more about the Malthusian argument that Jones uses as one of the pillars of his explanation for Asia's supposed lack of progress. The basic argument is, as we have seen, the irrationality of Asians. Asians do not plan their reproductive behavior. In China, the "energies of the peasantry diverted themselves away from higher consumption or even revolt into [settling] new land and breeding new people" (p. 219). Asians, in a word, accumulate children instead of capital. All of this is false. Some of it is bad theory: overpopulation explains rather little in Asian history, and did not really exist in most regions and epochs. Some of it is ignorance of the available evidence: Asians control their demographic behavior as rationally as Europeans do; they have practiced birth control for millennia. And finally, it is truly bizarre to state that in Asia "copulation was preferred above commodities."

GROWTH RECURRING

The argument of *The European Miracle* is modified and softened in *Growth Recurring*. Some of the changes have been indicated in the preceding paragraphs. It remains to summarize the essential positions that Jones takes in the later book.

The European Miracle argued a deterministic position: Asia and Africa could not have developed as Europe did. In Africa "it is not clear what indigenous developments were possible" (p. 156). In Asia, development "would have been supermiraculous" (p. 238). *Growth Recurring* backs away from this extreme position. In the earlier book Jones had stated his underlying theory about economic change and development: humans would change and progress if it were not for barriers, external to their bodies and minds, that prevent them from doing so. But this argument is obscured in *The European Miracle* by the abundance of derogatory statements about non-Europeans; Jones seems indeed to be saying that non-Europeans, as individuals, truly lack the qualities needed for economic progress.

The later book develops the basic theoretical argument more fully and modifies it in important ways. To begin with, pejorative statements no longer refer to Asians and Africans in general; they are directed mainly at the non-European political elites, whose attitudes and behavior Jones considers to be partly responsible for non-Europe's nondevelopment, and secondarily at non-European peasants, who, he believes, lacked the ability that Europeans had to control their reproductive behavior and thus were partly responsible for nondevelopment. Jones now makes a serious, though unsuccessful, effort to explain both maladies in terms of barriers external to the individual actors. Perhaps Asian rulers were despotic because they were heirs to Mongol conquerors; I pointed out above that this argument is illogical because the despotism is supposed to go back historically beyond the Mongol period and because you can't really blame the Mongol invasions for politics many hundreds of years later. Perhaps peasants had too many babies as a rational response to natural disasters and the depredations of their rulers; but here again the logic is faulty and the evidence is missing: natural disasters were not at all common and probably had no greater impact on individuals in non-

Europe than they did in Europe; and we know that at most times and in most places the peasants were not oppressed by the rulers to the extent that they could not retain enough product for their basic needs. The natural environment is also invoked again, but as a kind of generalized barrier, unexplained.

Probably the most important stand-down in *Growth Recurring* is the abandonment of the idea that *only* Europe progressed in the premodern epochs. But again the improvement is limited. Jones seems merely to be catching up with two doctrines that are now quite conventional: Song China was progressive for a time, and Japan had at least the potential for development as far back as the Tokugawa period. Jones offers no persuasive explanation for these two exceptions to the rule of non-European nondevelopment. Yet it is an important step forward for him to accept the proposition that under certain circumstances—even if the circumstances are not explained—development was not a European monopoly.

The conclusion that Jones reaches in *Growth Recurring* is that *some* progress occurred in Asia (though not, it would appear, in Africa[35]). But it was a limited sort of progress that he describes as "extensive growth." This means that technological and economic progress occurred just to the extent that it managed to keep pace with population growth—hence, no real change, a Malthusian horse race. What Jones calls "intensive growth" is *real* progress, fundamental change. This occurred only in Europe, Japan, and in the long-ago Song period in China. I do not see that we gain very much enlightenment by using these two labels, "extensive growth" and "intensive growth." However, the notion of "extensive growth" conveys an important step forward in Jones's thinking: even the attainment of "extensive growth" demonstrates that human beings naturally strive for economic advancement and achieve it at least to some extent. The barriers are not absolutely unbreachable.

NOTES

1. See Rostow, *The Stages of Economic Growth* (1960); Black, *The Dynamics of Modernization: A Study in Comparative History* (1966).

2. Jones, *The European Miracle: Environments, Economies, and Geopolitics in the History of Europe and Asia* (1981).

3. For example, in 1985 a symposium with the title "The European Miracle" was held at Cambridge University; without criticizing Jones, the contributors affirmed his basic argument about Europe's historical superiority but emphasized the intellectual and cultural causes, not (as Jones had done) the economic and environmental ones. See the proceedings volume: Baechler, Hall, and Mann, eds., *Europe and the Rise of Capitalism* (1988).

4. The page numbers in parentheses in this chapter refer to pages in *The European Miracle*.

5. But Jones does not really *admit* that assertions in *The European Miracle* had been wrong. He states some contradictory theory, giving the impression in some cases that it represents newer scholarship: "The view is emerging in the literature that . . . " (p. 135). Or he restates the original theory as speculation justified by the absence of contrary evidence: "given the paucity and vague form of most historical evidence it is hard to support any particular idea" (p. 120). "The problem . . . remains with us because the sources are almost intractable and the necessary research has not been done" (p. 142). Or he hedges: "The likelihood was very great from what we know . . . " (p. 150); "stark pictures are misleading, but . . . " (p. 131); and so on.

6. Jones also claims that ordinary people were more well-off in early Europe than in early Asia and that income was more evenly distributed in Europe. These assertions are unsupported by evidence, and they are false. See Frank, *ReORIENT* (1998); Twitchett and Mote, *The Ming Dynasty* (1998); Subrahmanyam, *Merchants, Markets, and the State in Early Modern India* (1990) Pomeranz; *The Great Divergence: China, Europe, and the Making of the Modern World Economy* (2000); Wong, "Political Economies of Agrarian and Merchant Empires Compared: Miracles, Myths, Problems, Prospects" (1999); Goldstone, "Colonizing History: The West Is Best Can't Pass the Test" (1999).

7. See, for example, Collins and Roberts, *The Capacity for Work in the Tropics* (1988).

8. Giblin, "Trypanosomiasis Control in African History: An Evaded Issue?" (1990); Turshen, "Population Growth and the Deterioration of Health: Mainland Tanzania, 1920–1960" (1987); Porter and Sheppard, *A World of Difference: Society, Nature, Development* (1998).

9. Harrison, "The Curse of the Tropics" (1979).

10. Russell, *Man, Nature and Society* (1967).

11. Note this comment by Burton Stein: "At no time in south Indian history until the nineteenth century is there evidence that the creation and maintenance of irrigation works was other than a local reponsibility" ("South India: Some General Considerations of the Region and Its Early History," 1982).

12. Taeuber, "The Families of Chinese Farmers" (1970), pp. 63–86; Hilton, "Individualism and the English Peasantry" (1980); Kertzer, "The Joint Family Household Revisited: Demographic Constraints and Household Complexity in the European Past" (1989); Berkner, "The Use and Misuse of Census Data for the Historical Analysis of Family Structures" (1975); S. Guha, "Household Size and Household Structure in Western India c.1700–1950: Beginning an Exploration" (1998). G. Lee, in "Comparative Perspectives" (1987), p. 65, points out that "[many] scholars contend that the majority of families in any society are and always have been nuclear, regardless of the cultural elements favoring extended families." Also see Goody, *The East in the West* (1996).

13. F. Hassan points out that "the practice of population control in one form or another is universal"; in "Demographic Archeology" (1978), p. 71. Also see Nag, "How Modernization Can Also Increase Fertility" (1980); Pomeranz, "From 'Early Modern' to 'Modern' and Back Again: Levels, Trends, and Economic Transformation in 18th–19th Century Eurasia" (1999a).

14. Jones claims, for instance, that the true rural dwelling house and the chimney are medieval European inventions, giving no evidence for this peculiar suggestion. Like Lynn White (Chapter 3), he then attributes marvelous effects to these inventions: "a

controlled microclimate," respect for "individual distance," lower infant mortality rates, and more.

15. Also: "In most instances the origins of these techniques are misty; elaborations and even independent discoveries within Europe need to be considered. When individual inventions are looked at closely they often prove not to have been single formative events . . . but accretions and improvements. . . . European society could generate novelties and it was capable of borrowing effectively" (p. 57).

16. Needham, *Gunpowder as the Fourth Estate East and West* (1985).

17. See Needham, *Science and Civilization in China, Vol. 4, Part 2: Physics and Physical Technology: Mechanical Engineering*(1965) on Chinese clocks and mechanical engineering.

18. See, for instance, Hucker, "Ming Government" (1998); Goody, *The East in the West* (1996); Frank, *ReORIENT*; Perlin, *The Invisible City: Monetary, Administrative and Popular Infrastructures in Asia and Europe, 1500–1900* (1993); Arasaratnam, *Maritime India in the Seventeenth Century* (1996); Heijdra, "The Socioeconomic Development of Rural China During the Ming" (1998); Rowe, *Hankow: Commerce and Society in a Chinese City, 1769–1889* (1984); Marks, *Tigers, Rice, Silk, and Silt: Environment and Economy in Late Imperial South China* (1998); Yang Lien-sheng, "Government Control of Urban Merchants in Traditional China" (1970); S. Mann, *Local Merchants and the Chinese Bureaucracy, 1750–1950* (1987); Abu-Lughod, *Before European Hegemony: The World System A.D. 1250–1350* (1989).

19. Pires, *The Suma Oriental* (1944), pp. 41–42.

20. See *The Colonizer's Model*, Volume 1.

21. Blaut, *The National Question: Decolonizing the Theory of Nationalism* (1987), Chapter 7.

22. See Pounds, *An Historical Geography of Europe* (1990).

23. See, for example, Kumar, "Private Property in Asia"; Abu-Lughod, *Before European Hegemony* (1989); Hucker, "Ming Government"; Frank, *ReORIENT*.

24. For instance, for ancient India, see Mukerji, *The Republican Trend in Ancient India* (1969); for early-modern Morocco, see Abun Nasr, *A History of the Maghrib* (1975), p. 218.

25. Jones, *Growth Recurring: Economic Change in World History* (1988), p. 127.

26. Jones, *Growth Recurring*, p. 8. But the argument is hedged: "On a time-scale of a century or so [after Mongol rule ended] the economies recovered. . . . This is sufficient to cast doubt on the impression that even the most destructive of all the invading hordes can be blamed for a long-lasting diversion of whatever original prospects for [intensive economic] growth there had been" (p. 111).

27. We should note that commensalism is a biological term referring to a form of mutualism that may exist between or among animal species; I have not previously seen it applied to humans.

28. See, for example, Hopkins, *An Economic History of West Africa* (1973), pp. 36–37.

29. Here is another example of Jones's ignorance of geography, the science upon which he claims to rest so much of his case. In drier areas, says Jones, "soils are ancient and poor, having been leached to the poverty line" (p. 154). In actuality, leaching is not significant in soils of drier areas. Some African soils are poor, some are rich, some are ancient, others new.

30. Shifting agriculture, or forest–fallow rotations—burning of forest, followed by

planting, followed by long fallow periods during which the forest regrows—does *not* lead to degradation of the environment, as Jones claims. He concedes that shifting agriculture was "efficient" when population density was low, although even here it was not "productive" and "land was not given long enough to rest" (p. 154). This is another old colonialist myth designed to show how irrational were the practices of native peoples. (See Blaut, "The Nature and Effects of Shifting Agriculture" (1962) and "The Ecology of Tropical Farming Systems" (1963); Nye and Greenland, *The Soil Under Shifting Agriculture* (1960).

31. This was one of the key myths of *apartheid*: that whites do not have any obligation to return the land to Africans because they, the whites, arrived in South Africa before the "Bantus" did. This is completely untrue. Jones introduces it, apparently, to show that Africans were so backward that they had not even discovered all of Africa before the Europeans did. (See Chapter 8.)

32. Jones invokes peculiar population figures for prehistoric and historic Africa to show that the population was low in comparison to world population and grew proportionately lower. He cites no authority. He asserts that the population of Africa was 30 percent of the world's population in 10,000 B.C. But we know almost nothing about population levels in 10,000 B.C. He then says "in A.D. 500 it was already down to ten percent of the world's people and a thousand years after that was about the same; by 1800 that proportion had shrunk to a mere eight percent" (p. 155). Again: nobody has a clue as to accurate population figures for A.D. 500. In A.D. 1500, according to C. Clark (1977), Africa had twenty percent of the world's population (eighteen percent, according to Bennett, 1954), not ten percent. In 1800 Africa's population was indeed low (twelve percent of the world's total)—because of the depredations of the slave trade.

33. There is little of this language in *Growth Recurring*, but the theory persists: "China seems not to have produced a sharp-edged, experimental approach of the type that really may lead to better technologies" (p. 75).

34. As a geographer, I am particularly distressed by this falsehood.

35. "One may . . . confess to a sneaking sympathy with Trevor-Roper's view that the main function of African history is to show the present the face of the past from which it has escaped" (Jones, *Growth Recurring*, p. 90).

CHAPTER 6

Michael Mann:
The March of History

M ichael Mann, in an essay titled "European Development: Approaching a Historical Explanation," asserts that a "miracle" occurred "spontaneously" in ancient and medieval Europe. He objects to what he calls "European self-denigration," the attitude of historians who believe that Europe was *not* superior to other civilizations prior to the late Middle Ages.[1] In particular he criticizes Joseph Needham, the great historian of Chinese science and technology, for arguing that China was on a par with Europe until at least 1450. Mann presented the argument in his widely discussed book *The Sources of Social Power, Volume I: A History of Power from the Beginning to A.D. 1760*; and he argued further in the later essay, "European Development," that Europe was always superior, in countless ways: its superiority first appeared in prehistory and thereafter become reinforced with new superior cultural qualities in succeeding epochs down to the present.[2] According to Mann, the "European self-denigration" of historians like Needham results from their failure to see that what Asian civilizations had was not *better*—it was simply *larger*. Or, as Mann expresses it, Asians developed "extensively," Europeans "intensively." Thus, for example, medieval Asian empires were vast in extent, their riches much beyond anything to be found in Europe, their cities larger, and so on, but this was merely an "extensive" matter. What does "intensive" mean in this context? It means: better.

"THE LEADING EDGE"

Mann's argument has a distinctive geographic quality. He believes that cultural superiority traced a steady course northwestward from the Middle East to Greece to western Europe, on a path that reminds us of the route of the Orient Express traveling on the westbound track. We might think of it as the Occident Express.

> Over several millennia there had been a drift of "the leading edge" of power in the Near Eastern/Mediterranean/European culture area to the west and north. (p. 17)

"Power" here means, basically, the level of civilization as expressed in many cultural achievements, including technological and economic productivity, military strength, and the complexity of social and political organization.[3] Mann believes that the "leading edge" of all of this moved inexorably northwestward (or west-northwestward). This happened, he says, for the following macro-geographic reasons.

Mann starts where Eric Jones started in his book *The European Miracle*: with a distinction between the irrigated agriculture of Mesopotamia and the unirrigated, rain-fed agriculture of Europe. Mann believes that Mesopotamia and other civilizations of this type, including Egypt, were locked into a nondevelopment trap because of the classic Oriental despotism syndrome, descending from the nature of irrigated agriculture.[4] (We discussed the theory of Oriental despotism in Chapters 2 and 5.) Quite distinct from these civilizations were the crude yet progressive societies that Mann calls the "Iron Age peasants" of Europe. He argues that rain-fed agriculture plus the acquisition of the iron plow produced a society of free and energetic and individualistic and commerce-minded agricultural folk, the true root of Europe's progressiveness, the Indo-European-speaking Greek and Germanic peoples who acquired some elements of civilization from the ancient civilizations of the Middle East—later from the Roman Empire—and thereafter, because they were not hampered by Oriental despotism and the rest, forged ahead toward modernity. Here is how Mann expresses the theory, in a chapter of *The Sources of Social Power* titled " 'Indo-Europeans' and Iron: Expanding, Diversified Power Networks":

> Iron's cheapness meant that . . . [settled] agriculture, rain watered and not dependent on artificial irrigation, was boosted, and the peasant farmer grew as an economic and social power. The balance of power shifted from pasto-

ralists and irrigating agriculturists to the peasants of rain-watered soils. . . .
from aristocracies to peasantries . . . In geopolitical terms economic growth
shifted disproportionately toward the lighter rain-watered lands of
Anatolia, Assyria, southeastern Europe, and the northern Mediterranean.
This region developed an economy in which the individual peasant household related directly to . . . exchange and . . . specialization . . . —a boost to
private small-scale property and to the democratization and decentraliza-
tion of economic power.[5]

The entire construct is fallacious. (Most of it comes from Max
Weber.[6]) Iron plows were used in Asia as well as, and perhaps earlier
than, in "Iron-Age Europe," so the consequences were not peculiar to
"Iron-Age Europe."[7] Rain-fed agriculture was important in most of the
ancient civilizations (for instance, India, Persia), dominant in some regions (for instance, North China, Anatolia), and insignificant in only a
few civilizations (most notably Egypt, where rainfall was very low); irrigation was used in some parts of southeastern Europe in the first millennium B.C.; so Mann's model of a sturdy, independent farmer, not dependent on any power center for the water needed by his crops, and for
this reason an incipient democrat and entrepreneur, must apply to most
of the Eastern Hemisphere if it applies anywhere (which it does not:
lack of evidence). The idea that a society of sturdy yeoman-type, independent farmers emerged *uniquely* in Europe is simply a myth, and the
arguments from geography—rainfall, iron technology, and so on—are
not good science.

   We should note some additional problems with this theory. First, the
model of progressive society is attached by Mann to "Indo-Europeans,"
using a theory about the role of "Indo-Europeans" or "Aryans" in history
that is now much disputed, partly on evidential grounds, partly on
grounds that the term retains the aura of past prejudice.[8] Second, in line
with his Occident Express model of northwestward historical movement,
and his notion that "Indo-Europeans" were somehow central to the new
peasant individualism, Mann introduces a peculiar theory about the role
of the Greeks in the process. They became an important station on the
Express apparently for two reasons: they were the southernmost Iron-
Age, Indo-European-speaking peasants, and they lay at the boundary between the land civilizations of the Middle East and the Mediterranean
trading area. Phoenicians are dismissed on the traditional false ground
that they were somehow not democratic like the Greeks (because they
were not Indo-European-speakers? because of their areal proximity to the
Middle Eastern despotisms?). In any event, Greece is an important station on the "Occident Express." Greece is credited by Mann, in the tradi-

tional manner, with providing history with democracy, a "multi-state system," literacy, "confidence in reason," awareness of the laws of nature, and so on.[9]

> Athens saw probably world history's most genuine participatory democracy among an extensive citizenry (still, of course, a minority of the whole population—for women, slaves, and resident foreigners were excluded).[10]

Some have noted that a society in which the majority are slaves is not a democracy, no matter how much equality there is among its elite minority (of males). Mann fails to take notice of contemporary oligarchy–democracy experiments elsewhere (for example, in northern India[11]), of literacy outside of Greece, and of science in other societies—Egypt was, if anything, well ahead of Greece in abstract science and mathematics.[12] Given that iron was used in China and, indeed, throughout much of the hemisphere at the same time it was used in Europe, and given that plowing was also widespread, in irrigated agriculture as well as rain-fed agriculture, and given the rest of its difficulties, Mann's theory about the origins of European progressiveness does not convince. It is bad history and bad geography.

"RATIONAL RESTLESSNESS"

Mann now brings us into the Christian era. Christianity is the gift to Europe of the ancient Middle East, transmitted through Roman society, and, farther north, melds with Europe's root society, the bumptious, independent, competitive, democratic, restless, rational society of individualistic peasant farmers: the Indo-European "Iron Age peasants." This produces, quickly and (in my eyes) magically a single, definite, organic European society with a genuinely teleological goal: to push its "leading edge" ever farther toward the northwest and in so doing to advance and develop.

All that Mann says by way of explaining this northwestward push of Europe's "leading edge" is to offer us, first, the mystical image of Europe as a unitary, telic social organism, then, second, to list what he calls the "blockages" to growth eastward and the "opportunities" for growth northwestward. The "blockages" to the east are caused by the presence there of strong (inferentially, Oriental despotic) states and barbarian hordes. The very idea of an entity, Europe, being blocked from growth in one direction by another entity, described by Mann as the Oriental empires and then as the entity "Islam," is highly metaphysical as well as historically telescopic. It evokes for us the idea that Dark Age Europe acted as a single social entity and so did its "blockers" to the East. One has to penetrate deeply into Mann's text to realize that he credits Christianity with having created, very early and very thoroughly, a single, essentially organic decision-making entity, deciding which way Europe can grow and which way it is "blocked" from growing. We return to the role of Christianity in a moment.

Mann introduces a subordinate argument to explain why development did not take place in Italy and other regions of southern and southeastern Europe, along with Byzantium. These countries had to "defend the eastern frontier," and in so doing must "drain themselves." Thereafter they "would be unlikely to make a major contribution to the European dynamic" (p. 17). Again we have the notion of an organic Europe, with southern and eastern Europeans serving as a kind of defensive rear guard. But as an explanation for the nonparticipation of southern and eastern Europe in the putative medieval social revolution, with its northwestward "leading edge," this is very thin stuff. Most models of cultural evolution tend to see the greatest progress in areas where ideas and culture traits are mixed and borrowing is greatest, particularly in matters military. (Mann places considerable weight on military culture, particularly in *The Sources of Social Power*.)

Blocked to the east, Europe therefore sees its "opportunities" to the west. Now Mann plugs into his theory a combination of Jones's environmentalism (see Chapter 5) and Lynn White's technological determinism (see Chapter 3). Mann takes from Jones the mistaken idea that the soils of northwestern Europe have an inherent fertility, an inherent potentiality for highly intensive production, that is unmatched elsewhere, and he builds for us the image of Dark Age and early-medieval northwest-European society exploding into economic growth as it begins to exploit these marvelous soils. Actually, as we saw in Chapter 5, this environment, overall, has moderately good productive potential, but to make these sorts of exaggerated historical claims for it is quite absurd. Staples such as wheat in premodern times grew moderately well in some of these soils, but the yields were held down by the overabundance of rainfall, which not only saturated the soil but brought with it a lot of cloudiness and so diminished sunlight and therefore photosynthesis; yields were held down also by the acidic and often nutrient-poor status of these mainly podzolic and gley soils. And a vast swathe of land across the North European Plain was not at all productive until the arrival of the potato from South America. (The potato thrives in moist, cool environments.). The soils of northwestern Europe had considerably *lower* fertility potential than the soils of many other farming regions. If we take the most technologically advanced soil-management practices that were in use in each of these regions around one thousand years ago, the potential yield of staple grains was generally greater on many classes of non-European soils (such as some of the loess soils of northeastern China, the volcanic soils of Java, the black cotton soils of India, the alluvial soils of many river basins, and of course the soils under irrigated wheat and rice) than on the best soils of rainy northwestern Europe.[13] Mann's theory is therefore invalid.

Early in the Middle Ages, northwest Europeans carried out a technological revolution on these marvelous souls. The position here is taken directly from Lynn White and repeats all of the exaggerations and errors which we noted in our discussion of White's views in Chapter 3.[14] There were four crucial inventions, says Mann, that "probably gave western Europe a decisive agricultural edge over Asia, particularly over Chinese intensive rice cultivation techniques" (p. 11). No evidence for this strange assertion is given or indeed exists. Intensive wet-rice cultivation in China, in that period, gave much higher yields than any medieval agricultural system in Europe. The four inventions, per White, were: "plowing, shoeing and harnessing of draught animals, field rotation, and the water-mill" (p. 8).[15] Mann does not notice that the heavy plow was in use in India a millennium earlier, harnessing and shoeing technology was developed over much of the Old World, but particularly in central Asia, intensive rotations and rotations with stubble grazing were in use in many non-European areas, and the water mill is recorded from China just as early as it is from Europe.[16] Mann quotes White as to the effects: a "novel system of agriculture" (not true unless we stretch our definition of "novel": see Chapters 3 and 9); the acquisition of "intensive social organization" and cooperative activity in the villages of northwestern Europe (an error that we discussed in Chapter 3).

Mann proposes to "quantify" this "early medieval dynamism" by noting that there was an increase in population and in the yield ratios of crops. Population did indeed grow, but there is no reason to attribute it to an agricultural revolution. Indeed agricultural development, particularly in the assarting of new lands, had much to do with it, but the coming of general peace was perhaps of greater significance. And we are obliged to ask why population growth should be considered a positive sign of development in medieval Europe but a Malthusian curse elsewhere. And Mann's reference to increasing "yield ratios" is also erroneous. Yields as such did indeed increase, but the degree to which this reflected technological change was minor, as compared with such things as the extension of cultivation to new and fresh land.[17]

Mann wants to demonstrate that rapid technological, economic, and social development was occurring in northwestern Europe deep in the Middle Ages, in contrast to the rest of the world, and that this reflected a very dynamic society, again in contrast to the rest of the world. It is unquestionably true that change was taking place in this region and era, but the major thesis is unquestionably *untrue*: change was not revolutionary, a true beginning of Europe's "take-off" into industry and modernity, and comparable change was indeed occurring in many regions outside of Europe. (See the discussion of David Landes's theory of a medieval technological revolution theory, in Chapter 9.) So, Mann's factual thesis—unilateral medieval progress in northwestern Europe—is simply wrong. Let us now look at the way he explains this revolution that did not in fact occur.

At this point in the argument, Mann brings in, very crucially, Max Weber. These northwest European people were uniquely "rational" and "restless." They displayed the incipient qualities of modern capitalist man: individualism, dynamism, and so on. Why did they display these qualities? Mann first introduces his thesis about the old "Iron-Age peasants" who bequeathed some of these qualities to their descendants, the Germanic and Celtic peoples of northwestern Europe. We are given the image of peasant life deriving from the free-living Germanic tribes, while manor life, with its feudal knights, inequality, serfdom, and the rest, comes from a vaguely "un-European" and "Eastern" source, via Rome.[18]

But, for Mann, the crucial element in explaining medieval dynamism, rationality, restlessness, and the rest of the qualities that pushed the "leading edge" of Europe northwestward and generated rapid development, is Christianity. Christianity integrated Europe into a single society. (Indeed this is true although the degree of unity was much less than Mann claims.) Christianity, according to Mann, imbued Europeans with many of the character traits needed for the technological and economic revolution that (he says) they were about to carry out. It provided a set of norms, so that Europeans would "trust one another to honour their word" and "trust each other's essential rationality" (p. 11). Beyond that, Christianity gave Europeans the quality of "rational restlessness." These admirable character traits were actually "strands of the Christian psyche which were traditionally present." One such strand was "ethical individual conduct." (Most of this comes directly from Max Weber, and clearly implies, as did Weber, that non-Europeans lacked these traits.) Christianity "encouraged a drive for moral and social improvement even against worldly authority" (p. 12). This extraordinary statement perhaps most fully reveals how Mann departs from traditional interpretations of the social role of medieval Christianity, as a basically conservative force, generally supporting authority, looking historically backward to the Fall, conceiving the universe as a "great chain of being," preaching personal salvation not social change, and often blending with political authority and defending the status quo. Indeed, Christianity did provide a common set of ethical values and much of a common culture to Europeans, but so did other great religions to the people of other regions. Perhaps medieval Christianity, and the Church, was, on balance, neutral in the political and class struggles of the time, and certainly it did contribute to the forward march of social and material progress. To say, however, that Europeans were uniquely progressive, dynamic, rational, trustworthy, and the rest because of Christianity is very unconvincing—another manufactured explanation for the superiority of Europe and Europeans.

Mann gives Christianity another crucial role in Europe's supposedly unique medieval progress. Often enough it is argued that the period of feudal disunity was detrimental to European progress and that an acceleration of progress began when political unity, finally, was reestablished. (That there *was* progress during the Middle Ages in Europe is not at all in dispute. The point is that progress was *also* occurring in the other continents: Europe was not unique, as Mann insists it was.) Mann wants to argue, in quite the opposite way, that the disunity of feudal Europe was actually a cause of its progress. This argument (which is also made by Jones, Hall, and Landes) is in part a return to the theory of Oriental despotism and the correlative argument that Oriental empires, because they *were* empires, despotically stifled progress and maintained the status quo of backwardness. (I will comment further on this political theory when we turn to the views of John Hall in the next chapter.) Says Mann: Europe escaped the stifling of progress in part because Europe did not have political unity. It was, instead, "a multiple acephalous federation," with "no head" and "no centre" (p. 11). (I am not sure where "federation" comes in since there was no political federation of significance in the period under discussion.) Mann appears to anthropomorphize the feudal fragmented polities: he sees them as being competitive, individualistic, vigorous, etc., in their behavior toward one another, much as he sees medieval people as having these qualities, and he infers a kind of implicitly capitalist personality to both. Christianity, he argues, substituted for political unity in the ways in which political unity was positive, not negative. (Recall a similar argument put forward by Jones.) Thus Europe had the best of both worlds.

There is, finally, one additional environmental "opportunity" that, so to speak, puts the finishing flourishes on the superiority of Europe and its unique rise. This, says Mann, is the coastline of western Europe. It is indented and thus favors seafaring and sea trade. It is westward-facing, and so beckons Europeans to venture across the sea. This is in fact an old theory, partly environmentalistic, partly mystical. Europeans, in this theory, have always had the urge to expand and have succumbed to this urge in all epochs, from the Crusading days on down. Facing the sea, Europeans feel impelled to cross it. Significantly, Mann insists on the antiquity of this urge, and this permits him to avoid having to give causal efficacy to the precise epoch of Iberian exploration and the conquest of the New World, which become simply episodes in a much older and deeper dynamic. We discussed this theory in the preceding chapter; suffice it to say now that exploration and expansion were occurring during this period in other maritime–mercantile societies of the late Middle Ages: many of them had this "urge to expand."

We can now, I think, assess the theory as a whole. Mann believes, with Jones, that Europe, uniquely among world civilizations, possessed the qualities needed to rise and to develop a capitalist economy, and also was vigorously moving forward toward capitalism, deep in the Middle Ages. He believes that the environment of northwestern Europe was uniquely suited to technological and thus economic advance and that the people of the region were heirs to a post-Neolithic culture that was not very far removed from capitalist culture in its individualism, freedom, competitiveness, and the rest. He sees all of this happening in local communities of rural Europe, communities that were anything but class-ridden:

> The image is of small groups of peasants and lords standing looking at their fields, tools and animals, figuring out how to improve them, with their backs to the world. (p. 5)

Thus: a brotherhood of serfs and their masters.

However much environmentalism, technological determinism, and cultural–historical Eurocentrism is present in Mann's theory, Christianity is, after all, the central part of the explanation. It enables the northwest Europeans to perform their technological magic, to create trade, markets, and the preconditions for a capitalist economy, and to perfect, as it were, the Weberian personality type which had only been rudely carved out in the prehistoric days of "Iron-Age peasantry." It provides

Europe with a sharply-defined cultural identity. Mann, like many Eurocentric historians, wants to add up all of the causes, or factors, that point toward European superiority; to, in essence, allow each of them to play some role in the process; and then to cap off the argument with his own preferred candidate: in this case, the "rational restlessness" created by Europe's religion and northern Europe's tribal origins. A particularly troublesome feature of Mann's theory is his lack of specificity as to *how* Christianity plays the role he gives it. He shows no awareness that the things he attributes to this religion can also, in varying ways, be attributed to other religions, in other cultures.

And where was all of this leading?

At the end of all these processes stood one medium-sized, wet-soil island state, perfectly situated . . . for take-off: Great Britain.

WESTWARD HO!

"The geographical march of history," said the geographer Arnold Guyot in the mid-nineteenth century, is an "incontestable fact."[19] Mann seems to agree: "development . . . was extraordinarily continuous and shifted steadily towards the north-west" (p. 10), "economic power continued slowly shifting north-west" (p. 16), "over several millennia there had been a drift of the leading edge of power . . . to the west and north," and so on. I will close this chapter with an effort to quantify Mann's conception of the westward or northwestward march of history. I will use the method, described in Chapter 4, of plotting dated place-name mentions (DPMs) on maps for successive historical intervals.

Figure 2 shows the distribution of dated place-name mentions in Mann's book *The Sources of Social Power, Volume I: A History of Power from the Beginning to A.D. 1760*, for all of history down to 1700 A.D., divided into 500-year segments (Figure 2a–g) with a terminal 200-year segment, 1500–1700 A.D. (Figure 2h).[20] I drew a sample by selecting the first DPM on each page of the book and plotted the resulting DPMs as dots on the appropriate maps. On each of the maps I plotted the median latitude/longitude of all place-name mentions for that interval.[21] The final map (Figure 2i) displays all of the median points for all of the historical intervals. Connecting these median points shows history's line of march as viewed by Mann.

The pattern is very clear. History, for Michael Mann, travels westward and northwestward, a sort of Occident Express.[22]

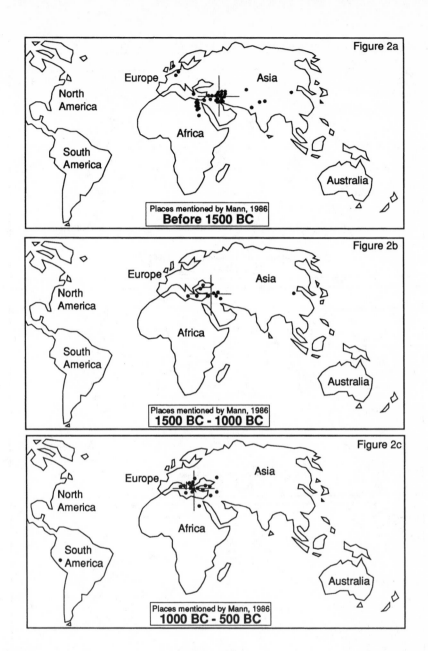

FIGURE 2. Dated place-name mentions (DPMs) in Mann, *The Sources of Social Power, Volume 1: A History of Power from the Beginning to 1760*, plotted on maps for 500-year intervals from before 1500 B.C. to 1500 A.D. (Figure 2a–g) and the 200-year interval from 1500–1700 (Figure 2h). (Sampling method: see text.) The crossing lines indicate the median latitude and longitude of all DPMs for a given historical interval. The final map (Figure 2i) shows the medians for the entire chain of historical intervals, with a line connecting the medians.

123

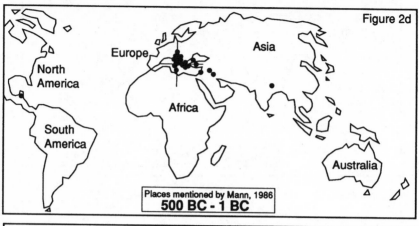

Figure 2d

Places mentioned by Mann, 1986
500 BC - 1 BC

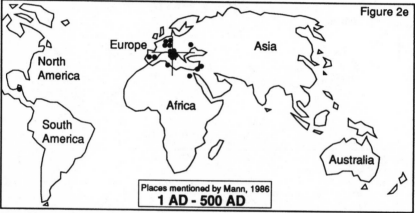

Figure 2e

Places mentioned by Mann, 1986
1 AD - 500 AD

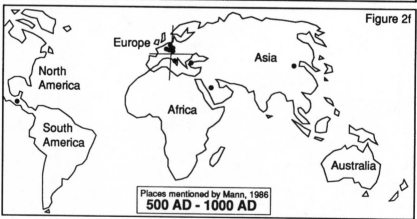

Figure 2f

Places mentioned by Mann, 1986
500 AD - 1000 AD

FIGURE 2. *(cont.)*

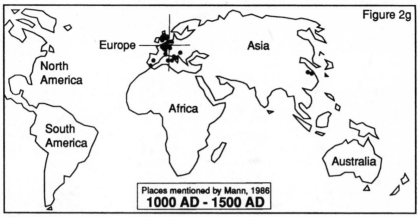

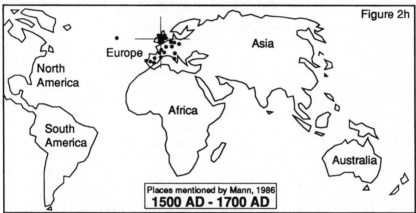

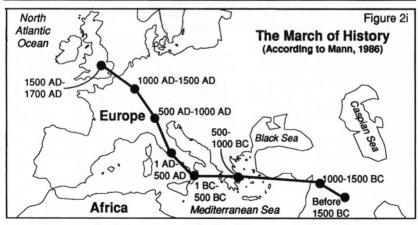

FIGURE 2. (cont.)

NOTES

1. M. Mann, "European Development: Approaching a Historical Explanation." In Baechler, Mann, and Hall, *Europe and the Rise of Capitalism* (1988), pp. 6–19; the words quoted above are on page 6. Page numbers given in parentheses in this chapter refer to this article; other works by Mann are cited in the notes.

2. M. Mann, *The Sources of Social Power: Vol. 1. A History of Power from the Beginning to A.D. 1760* (1986). Vol. 2 of this work, titled *The Rise of Classes and Nation-States, 1760-1914* (1993), is much more solidly grounded in evidence and theory. My main interest here is in Mann's theory of the march of history; this is expressed most clearly and succinctly in the article "European Development: Approaching a Historical Explanation," so I will refer mostly to that work.

3. The detailed argument is given in M. Mann, *Sources of Social Power, Vol. 1*. In this book Mann uses "power" to mean so many different cultural things that the work as a whole is in no real sense a history of "power" but seems rather to be a general social history. In a later volume, *Sources of Social Power, Vol. 2. The Rise of Classes and Nation-States, 1760–1914* (1993), Mann discusses different sorts of power (pp. 1–4), but the overall concept is very vague.

4. Mann's version of the Oriental despotism theory is similar to Weber's (see Chapter 2) but quite different from that of Karl Wittfogel (described in Wittfogel's 1957 book *Oriental Despotism*). Mann argues, in essence, that irrigation crowded ("caged") people together in Mesopotamian states of a despotic sort in which cultural unity was not achieved between ruler and people, the former primarily military and despotic. There is no recognition, in Mann's argument, that irrigation is not, historically, an independent variable. In most of the irrigated lands, irrigation was *selected* for social purposes, as when an elite class demanded increases in surplus production that could only be provided by developing or enlarging an irrigation system (see *The Colonizer's Model of the World*, Volume 1, Chapter 2). Causality should start with the society, and one should not claim that "irrigation" does this or does that. Mann adds an invalid argument to the effect that these barbaric societies could not sustain really large polities because of the difficulty of moving troops and resources over very great distances overland: mobile troops could temporarily hold larger areas, but they lacked the economic infrastructure for control except where transport was by water (see *Sources of Social Power, Vol. 1*, Chapters 3–5). This argument depends on a common fallacy. It is not true, as Mann maintains (as do also E. L. Jones and many others), that overland travel and haulage was limited to short distances (he thinks a maximum of 150 km); for instance, that a beast of burden carrying grain supposedly would have had to eat the entire load it carried to sustain itself over this distance. Mann fails to take into account the fact that beasts of burden graze by the roadside or trailside along the route of travel; the location of grazing areas strongly conditioned ancient routes of commerce (for example, the Inner Asian Silk Road, which in fact stretched for thousands of miles). Moreover, river transport upstream was very difficult in ancient times and open-sea boat transport was still crude and inefficient. Calculations of the relative cost of land and water transport for ancient times are speculative, and theories built upon such calculations are not sturdy.

5. M. Mann, *Sources of Social Power, Vol. 1*, p. 185.

6. See, for example, Weber, *The Agrarian Sociology of Ancient Civilizations* (1976), pp. 157–158. I discuss Weber's view in Volume 1, Chapter 2, in the section titled "Arid, Despotic Asia,"and in Chapter 2 above.

7. Bray, *Science and Civilization in China; Vol. 6. Part 2. Agriculture* (1984). I put

"Iron Age" in quotes because the term usually carries the meaning of a uniquely European historical stage; part of a linear evolutionary model.

8. See in this regard Bernal, *Black Athena: The Afroasiatic Roots of Classical Civilization: Vol. 1. The Fabrication of Ancient Greece* (1987).

9. M. Mann, *Sources of Social Power, Vol. 1*, p. 227.

10. M. Mann, *Sources of Social Power, Vol. 1*, p. 211.

11. Mukerji, *The Republican Trend in Ancient India* (1969).

12. On these matters, see, again, Bernal, who shows how nineteenth-century northern Europeans in essence *invented* the conception of ancient Greece as a European culture hearth in order to construct a history without non-Indo-Europeans (Jews, Phoenicians, Egyptians) and with as little influence of Latin Europe (Rome) as possible.

13. This generalization neglects local problems like salinity and overly intensive cultivation and local environmental peculiarities (soil texture, water table, and so on). It applies, however, to all of these regions, from northwest Europe to China and Southeast Asia.

14. Some of Mann's argument is taken also from the work of another technological determinist, Carlo Cipolla; see Cipolla, *Guns, Sails, and Empires: Technological Innovation and the Early Phase of European Expansion, 1400–1700* (1965).

15. Later in the discussion Mann corrects the conspicuous error: he means, not "plowing" but the heavy plow; not "field rotation" but the three-field rotation.

16. See the discussion of these inventions in Chapter 3. Mann concedes that the water mill was known to the Romans, but then, illogically, continues with his discussion about its "invention" in the Middle Ages and its supposedly marvelous effects just then.

17. "Yield ratios" are not the same as "yields," and the difference is often crucial, although Mann takes the first as surrogate for the second. Information about yield ratios (the ratio of seed sown to seed harvested) is fairly good for the Middle Ages, but little is known about actual yields. A low yield ratio can disguise high yields. For instance, 3:6 is a lower ratio than 1:3 (harvests respectively double the amount of seed sown and triple the amount sown) but gives more return per acre (three units as against two). Low yield ratios prevail on poorer soils, but more acreage can be cultivated or different rotations can be used and so higher total yields can be obtained. These problems, which Mann does not mention (either in the essay or in the book) make it very difficult to judge the effects of technological change in medieval agriculture.

18. M. Mann, "European Development," p. 16. If Mann had not wished to disparage the Roman and Oriental (imperial) contributions to medieval Europe, he might have said, simply: the Roman empire imposed a class structure on northwest European peasant society.

19. A. Guyot, *The Earth and Man* (1849).

20. The terminal year 1700 is used because world history textbooks generally expand their coverage to include the whole world, especially the colonial world, for modern times, and geographical patterns therefore become somewhat blurred. In this exercise I treat *The Sources of Social Power: Vol. 1. A History of Power from the Beginning to* A.D. *1760* as though it were a textbook in world history, which in a sense it is.

21. The median was used in preference to the mean because the latter statistic would give too much weight to very distant places. Thus: if—a hypothetical example—all but one DPM for a given historical interval were located in Europe but one DPM lay in China, the mean latitude/longitude would be to the east of Europe, and so would be less meaningful than the median, which would fall somewhere in Europe.

22. The Orient Express took slightly different routes in different epochs. The analogy should not, of course, be pushed too far.

CHAPTER 7

John A. Hall:
Democratic Europeans

J ohn A. Hall, following Adam Smith, believes that capitalism will tend to arise naturally in society if politics does not interfere with laissez-faire and if there exist no "blockages" to stop the process. Europe, he argues, evolved a political system during the Middle Ages which did not interfere with the rise of the market and other attributes of capitalism; Asian societies, on the other hand—Africa goes unnoticed—had "blockages" that prevented this political state-shaping process from proceeding in the proper, natural way. Hall presented this theory in a 1985 book, *Powers and Liberties: The Causes and Consequences of the Rise of the West*, and in a 1988 essay titled "States and Societies: The Miracle in Comparative Perspective."[1] I will focus on the 1988 essay (and page numbers in the text will refer to this essay), but I will discuss *Powers and Liberties* where appropriate.

Hall's theory is somewhat similar to that of Michael Mann (Chapter 6), but Hall's candidate for Principal Cause is politics. Like Mann, he builds his theory on a foundation consisting of some of Max Weber's arguments about rationality and religion (Chapter 2), and he makes abundant use also of technologistic arguments taken from Lynn White, Jr. (Chapter 3), and environmentalistic, economistic, Malthusian arguments borrowed from Eric L. Jones (Chapter 5): thus another eclectic theory of the "European miracle," intended to get maximum support from a multiplicity of Eurocentric arguments but with a kind of signature of its own: in this case, politics and the state.

Hall wants to call his method "comparative" ("The Miracle in Comparative Perspective"), but it is the kind of pseudocomparative approach that we encountered in Jones's work and will encounter again in Landes's work (Chapter 9). First one lists all of the reasons why Asian societies were, historically, inferior in all matters related to progress and civilization (Africa not warranting discussion); then one lists all of the reasons why Europe was superior; then one compares the two and pronounces Europe the winner. Hall proceeds systematically: first he dissects "Imperial China," then "The Land of the Brahmans" (his label for historic India), then "Islam and Pastoralism" (a title with an embedded theory), and finally we are told about the wonderful "Rise of Christian Europe." (These are chapter titles in *Powers and Liberties*.) Hall's causal theory is woven into the discussion throughout.

Hall's theory of politics is close to that of Adam Smith, as he freely acknowledges.[2] He agrees with Smith that the less political interference there is in the economy, the better things will be: laissez-faire equals progress. Of course, says Hall, there has to be a sort of minimalist political environment to permit the economy to function: to ensure peace, lawful behavior, freedom of commerce, and the like. But that's about all. Hall's basic theory of the rise of Europe, disentangled from various peripheral arguments (which we will discuss in a moment), asserts that Europe modernized and developed capitalism mainly because Europe had a political system that did not "block" the free working of the economy, so the economy naturally developed—naturally because development toward capitalism is a natural thing to expect if there is nothing around to "block" it. "Block" is the word Hall uses, as do many other theoreticians of the European "miracle," to convey the idea that development is natural unless something artificial stands in its way. Such "blocks" are ordinarily to be found throughout the non-European world but not in Europe—not, at least, in northwestern Europe. This Smithian logic is also present in Jones and, at least implicitly, in White and Mann; and, rather surprisingly, in the neo-Marxist theory of Brenner, who thinks that capitalism rose in England partly because the state there did not interfere in the economy as it did in France and other European countries. In Hall, this is the central notion. It leads him to try to explain why it was that the state in other civilizations—China, India, and Islam—did not permit the economy to develop and why the European state did. Among the various "blockages" he singles out Oriental religions and despotism as the principal underlying factors.

"IMPERIAL CHINA"

In China, says Hall, the form of the state an was empire. He invokes the same indictment of what he calls "the imperial form" of state as does

Jones. There is "arbitrariness," "scorn for human life" (p. 20), people "whipped or killed" at the "whim" of the emperors (p. 34), and so on; thus invocation of a sort of Fu Manchu version of Oriental despotism. Hall's discussion of China basically telescopes three thousand years of history into a single, timeless description, so we are given the impression that the Chinese state eternally preserved the barbarism of its most ancient form. (Indeed, Hall wants to compare China in general with "ancient Egypt," p. 113.) The picture he gives us of the European state, by contrast, does not dwell on atrocities and arbitrariness but rather on humane and democratic institutions, and of course is a picture of relatively recent forms. Recall that Jones uses the same device to give a false impression of Oriental society: compare relatively modern Europe with *very* old traits of the Orient, but leave the impression that those traits are timeless. For Hall, the device permits him to make grand generalizations about the nature of "empires," and then heave a literary sigh of relief and say how fortunate Europe was to escape the "imperial form"—had it not done so, development would not have occurred in Europe.

> Imagine what European history might have been like had the Roman empire somehow been reconstituted, or had any empire taken its place! (p. 135)

Hall does not supply evidence to support his claim that "empire" stifled economic development in China. He simply repeats the assertion over and over as though it were an absolute and unquestionable principle and then adds a few irrelevant illustrations.

Hall repeats Weber's argument (see Chapter 2) that Chinese cities were not autonomous from the polity and infers that this somehow meant that the urban economy was stifled by the imperial state, whereas in fact the urban economy of premodern China was always massive and vibrant, long-distance trade as well as local exchange was of immense importance, the merchant community was omnipresent and sometimes shared personnel with the imperial bureaucracy, and the state probably helped the economy more than it hindered it.[3]

Hall then brings in the case of Admiral Zheng He (Cheng Ho), who organized massive expeditions which went to India and Africa during the period 1417–1433 (we discuss this further in Chapters 8 and 9). For Eurocentric historians, the important thing about Zheng He's voyages is not that they took place—representing perhaps the greatest oceanic exploit in human history up to that time—but that they stopped. The Chinese government stopped sending out the great fleets in 1433. This, according to Hall, proves that the empire was hostile to development in all forms and more particularly was xenophobically hostile to intercourse with foreign nations. But, as Victor Purcell and many other sinologists have pointed out, the most significant thing to be explained is not that they eventually ended but rather that these great voyages did take place—a half-century before anything remotely comparable took place in Europe.[4] They had a political purpose: basically, to caution Asian kingdoms against opposition to China and to reinforce tributary relations with some of these kingdoms. That purpose was accomplished and required no further voyaging. Merchants in considerable numbers went along on the voyages, and so they were a major step forward in economic development. They ended because of internal political crises in China and military problems on the northwestern frontier, and it is indeed true that the government thereafter discouraged, even occasionally prohibited, private oceanic trade. But that trade continued at an intense level of activity, largely uninhibited by the restrictions. Hall believes, curiously, that Chinese actually forgot how to build ocean-going ships after 1430.[5] If, indeed, there is any need to speculate about the ending of Zheng He's expeditions in the context of comparing European and Chinese exploration in the fifteenth and sixteenth centuries, the critical point is altogether different: Zheng did not discover America; the Europeans did so (the reasons are discussed briefly in Chapter 1 and more fully in *The Colonizer's Model of the World*); it was mainly a matter of accessibility); hence, Zheng did not have the good fortune to open up an almost limitless source of wealth for his country, as Columbus did for his.

The picture of China that is painted by Hall is thoroughly distorted. It matters not that he wants to discuss mainly the political system and does not, therefore, have to give us a general history of China. He wants to show that China's imperial system blocked economic development, but to do this properly he would have to tell us something useful about, at the very least, the economy. This he does not do. Rural society goes essentially unmentioned. Rural classes and class struggles are brushed aside. A diagram of China's class structure (adapted from Ernest Gellner) actually fails to include the landlord class (p. 9). The statement is baldly made that classes and class struggle in China were not "developmental," as they were in Europe, and did not produce "social progress" (p. 44).

China, says Hall repeatedly, was overpopulated: simple population growth there becomes transmuted into "overpopulation," whereas in Europe population growth means progress. China, says Hall, had greater inequality than Europe: supposedly a steeper "slope of inequality." This claim, unsupported, and apparently borrowed from Jones, seems merely to be a twisting around of the fact that China's greater riches meant that the elite was more affluent than that of Europe, so that it might seem that there is a stronger contrast with the ordinary person's standard of living. In fact, the average standard of living in China during that period was probably higher than in Europe.[6] (None of this would suggest that the state suppressed the economy.) And, apart from a handful of false and rather uninformed statements about Chinese culture (Hall considers it, for instance, a "passive" culture), no argument is given to explain the patterns that Hall claims to see.[7] Occasionally Hall cites Weber, approvingly, but he does not invoke Weber's theory about the inimical effects of Confucianism. We are left with the notion that China simply has been a nasty imperial society, an Oriental despotism, perpetually going through cyclic changes and ending up where it started, for no real reason. But Hall's ideas about Oriental despotism, stagnation, a "dynastic cyclic pattern," a perpetual "stalemate and so on" (p. 21) can be rejected. And we can reject Hall's claim that China did not have the potential to modernize prior to the coming of the Europeans:

> [The] whole tenor of the [Chinese] social formation was one which ruled out the creation of any real dynamism. (p. 34)

The Europeans brought "progress" (p. 56).

It remains to notice one more curiosity about Hall's treatment of the Chinese empire. In his discussion of European development, which we will turn to in a moment, he concedes, as he must, that the state had certain crucial functions to perform in the rise of capitalism; that total absence of state involvement in the economy would certainly have prevented the rise of capitalism, because the state was needed to supply the social conditions (peace, for instance) that infant capitalism needed. But if China had a government powerful enough to *inhibit* the development of capitalism, does it not follow that this government would have been powerful enough to perform the functions needed to *permit* capitalism to develop? At this point Hall becomes very subtle. You can have a government that is both *too* strong and not strong *enough*. That was the case in China. The government was strong in its ability to inhibit capitalism but not strong in its *lack* of ability to provide an infrastructure for capitalism. This political schizophrenia, says Hall, is a characteristic of empires. They seem to be strong but they are really weak. This is not a "contradiction," says Hall, merely a "paradox" (p. 20). To me it is a verbal device designed to enable Hall to denounce "Oriental despotism" in China and praise the "strong state" in Europe.

"THE LAND OF THE BRAHMANS"

India fares even worse than China. For India, as for China, Hall puts forward the traditional European stereotype of the country and its deficiencies. Just as premodern China is traditionally viewed by Europeans as despotic, so India is stereotyped as a country ruled by caste rather than by politics or economics: these latter are considered, in this stereotype, to be gifts brought by the British.[8] This is the line that Hall takes. Indian history is almost literally reduced to this single trait, caste, as though no other forces were important and as though caste explains everything that happened in India for more than two thousand years. Hence the chapter title in *Powers and Liberties*: "The Land of the Brahmans."[9]

Hall's argument about India can be summarized very simply. More than half of his discussion is devoted to ancient times, the time of the formation of Hinduism and Buddhism, and most of the remainder of the discussion is devoted to demonstrating that these ancient patterns continued thereafter to control Indian society and prevent it from developing; prevent it even from acquiring the normal traits of a civilization. Thus another example of the practice of imputing some negative or positive trait to an ancient culture and claiming that this trait determines the character and conditions the fate of that culture forever after, giving it a tendency either toward dynamism (Europe) or toward stagnation (India).

According to Hall, Brahmanism and the caste system diverted social power from the political realm to that of religion and caste, and to the Brahmans as the power-wielding caste, from very early times. Thereafter, he says, India had no real politics and no real state. The Brahmans "blocked the emergence of powerful polities" (p. 27). "India did not have a political *history*."[10] He then paints a picture of India, throughout its subsequent history, as a country ruled by the Brahman caste and not by kings or emperors. Therefore politics in India was epiphenomenal and vaguely ludicrous. States did not last long. Kings were obligated only to fight presumably useless wars, and they were thieves ("they simply took what they could," p. 28). The states were "predatory and . . . completely incapable of providing social infrastructure" (p. 29). All of this is faithful to the Eurocentric stereotype of Indian history as a kaleidoscope of wars and political peregrinations, none of which means anything: permanent nonchange, stagnation, nondevelopment. Hall simply employs a theory to the effect that the strength of caste and, within it, the social power of the (supposedly) politically unambitious Brahmans (who supposedly organize all social life in a social, not political, sphere), explains this signature of

Indian history: no politics, and no change. In fact, none of this is true. Strong Indian states were the norm, not the exception.[11]

Hall is mesmerized by the fact that India was united under a single subcontinental empire for only three relatively brief periods in its history; this seems to be the basis for his argument that there was permanent political chaos. Interestingly, as we will see, he takes prides in Europe's ability to avoid the evils of imperial rule; in Europe, its absence is seen as progressive, whereas in India the lack of a universal empire is seen as the very opposite. India, be it noted, is a very large piece of territory; and the subcontinent had a very large population; and the difficulty of unifying it in one imperial state says nothing about the strength of its smaller states. Indeed, many of these were very long-lasting and, by the standards appropriate to any given historical epoch, were quite strong. Generally, there was one large state in control of much of the middle and lower Ganges throughout long eras, in spite of dynastic changes of the sort very familiar in Europe. In South India, the Vijayanagar state lasted for a millennium. And there were many smaller states. To ignore all this and claim that India was, in effect, stateless is mere ignorance. In fact, it is a resurrection of the very useful myth perpetrated by the British in colonial times to the effect that India, basically, had no polity other than the tottering and essentially meaningless late Mughal empire, so there was no legal or diplomatic obstacle to the creation of a British Indian empire: India, was a political vacuum into which the British poured real politics. Moreover, this myth allowed the British to claim that local state laws governing land ownership were invalid (because the state was not a *real* one), so the British could appropriate land at will once they had accomplished the ritual of substituting Queen Victoria for the Mughal emperor, thereby giving her technical ownership of the land. (In the theory of Oriental despotism, the monarch supposedly owns all the land.) The claim that Indian history was somehow apolitical before the British came is both false and colonialist mythology.

Caste, says Hall, prevented India from having the political infrastructure needed for economic development. Caste also prevented the emergence of other social characteristics that are necessary for economic development. Because of the caste system, says Hall (in a truly bizarre comment), India "had no sense of brotherhood."[12] It was a society "based on division rather than the possibility of shared experience." This adversely affected economic interaction among the people. Land ownership was unstable because of the political chaos, so "peasants had no reason to invest." Caste "proved debilitating to economic life" (p. 28). Caste held the society in a rigid grip of nonchange and so prevented economic development. Statements of this sort are doubtless true for a few periods and some regions, but roughly comparable statements can be made about the social hierarchies and social obstacles to economic change in early Europe. As to more recent epochs, Hall contents himself with a few stereotypically familiar comments about the Mughals (for example, the "bizarre extravagance" of their aristocracy, p. 83). He fails to mention the fact that one-third or so of India was Muslim, not Hindu, in early-modern times, and so this substantial minority was essentially separated from the caste system. He fails to mention the immense literature on the caste system itself that has shown, among many other things, that most of its more rigid aspects were late to arrive and perhaps were not permanent, that the system as a whole was both highly flexible and highly variable in space, and that most of economic life was carried on with very little interference from caste rules. For instance, 90 percent or so of Indians were peasants of different castes, doing common activities regardless of caste membership.[13] But much more important are the facts about the economic development that *did* take place in India. When the British arrived, in the seventeenth century, the industrial structure of India was more advanced than their own. Commerce was intense throughout the subcontinent. Indian merchants in the coastal cities were well organized and were indeed part of a mercantile network that stretched across India and across (and beyond) the Indian Ocean, and that was unrivaled in its level of development of business techniques, banking, and the like during the period when Europeans first arrived on India's shores.[14]

How does Hall explain the curious characteristics which he claims to find in India? Apart from one or two nods toward environmentalism (for instance, he thinks that "jungle" separated the north from the south and may thus have interfered with political unification, p. 68), the only approach we have to an explanation is a kind of intellectual and religious determinism very reminiscent of Max Weber.[15] This would be evident alone from Hall's single-minded use of caste as an explanation for everything else. The power of caste is simply treated as a given: Indians accept it because of the control that religion has over their minds. Christianity is contrasted here: it is "a more rational type of religion." Christianity, also, was more "involved in politics."[16] This may seem contradictory to Hall's way of analyzing China's history, but it is no more so in Hall than it is in Weber, since rationality (and particularly religious rationality) is a basic causal category for both Hall and Weber. Remember that Weber placed a lot of causal weight on religion in China, but he was not above calling the Chinese people irrational—even thieves and scoundrels.

"ISLAM AND PASTORALISM"

The Islamic world connotes, to traditional-minded Europeans, nomadic, tribal people wandering across trackless deserts, descending from time to time on settlements to rob and pillage or convert people forcibly to their strange, fanatical, desert religion. This fantasy is still widely accepted in historical theories about the reason why Europe modernized and Muslim societies did not. Hall's theory is no exception; in fact, it is absolutely traditional in its use of stereotypes. To begin with, "Islam and Pastoralism" is the title of the chapter on Islamic civilizations in *Powers and Liberties*, and the section of that chapter that discusses the Islamic religion is given the heading "Monotheism with a Tribal Face." The idea of tribalism and nomadism is woven throughout his discussion. Essentially the entire explanation for the nonrise of Muslim society is contained in this model: the society, deep down, is really just a horde of fanatical nomads.

Let us put this matter in perspective before we look at Hall's argument. The term "Islam" can quite properly be used in discussions of premodern times to describe the entire region in which the majority of people are Muslims, just as "Christendom" can be used for another region. But at the end of the Middle Ages, as today, this religious region encompassed an extraordinary diversity of civilizations. There were indeed pastoral peoples in the deserts and grasslands of northern Africa and southwest and central Asia, peoples who perhaps never were genuinely "nomadic"—that is, were not wanderers without definite territory—but were nonetheless somewhat mobile and were keepers of herds. But some parts of the Islamic realm were rainy farming regions, among the most productive and densely populated of the world: for instance, most of Java, Bengal, and part of South India and the Ganges valley. Other areas were irrigated drylands and also productive and densely populated: Egypt, the Tigris–Euphrates valley, parts of the Indus watershed, and so on. So the geographical category denoted by "Islam," while it embraces vast drylands of the Eastern Hemisphere, principally denotes, as a social category, settled agricultural peoples whose lives have nothing to do with deserts, pastoralism, nomadism, or anything of the kind. Probably there were more Muslims in India and Southeast Asia than in all of the deserts of the hemisphere.

What justification is there, then, for juxtaposing "Islam" and "Pastoralism"? The obvious answer is that the Islamic religion came out of the desert region of Arabia and was propagated in the early period by Arabs who proselytized by moving out into many cultural regions, initially by conquest, later by peaceful means. It is also true that the Koran makes reference to desert conditions—as do the Old and New Testaments. But in later centuries the ties to Arab culture became subordinated in most regions to the local cultures—Javanese, Indian, Swahili, and so on—which, in most cases, were communities of agriculturists and merchants, not pastoralists. If we come down to the seventeenth century, the time of the rise of capitalism, it still made sense to speak of "Islamic society," but only if the usage was as broad as the one that includes under "Christian society" the Christians of the Philippines, India, Ethiopia, and the Americas along with those of Europe.

The main reason why Hall and other Eurocentric historians characterize Islamic peoples in a way that emphasizes nomadism, pastoralism, physical mobility, warrior culture, and the rest is because it fits into a functionally quite useful stereotype that has been used by Europeans for centuries to categorize those Islamic peoples who have—to put it bluntly—caused problems for Europeans: the North African and West Asian societies of the Middle Ages, and the obstreperous peoples who, in modern times, acted so irrationally as to resist colonial subjugation: the "fuzzy-wuzzies" of the Sudan, the "Barbary pirates" and the "Rif tribes" of North Africa, the "bogey-men" (Buginese, putative pirates) of Indonesia, and so on. Perhaps more importantly, during the colonial period, Muslim peoples needed to be stereotyped in some way that paralleled the symbolic models used for Africans, Indians, Chinese, and the rest, stereotypes that would, as it were, explain why they are not rational enough to manage their own affairs without colonial supervision. And today the same stereotype is used to explain why they are not rational enough to progress and develop except under the domination of European countries (including the United States) and European-owned multinational corporations. For these reasons, and others, the stereotype of Islamic peoples as "warriors on horseback," as fanatic and irrational tribesmen, and the like is still functional and, in scholarly circles, still widely accepted. One of its functions, of course, is to support the idea that Muslims peoples could not, under any circumstances, have modernized on their own—to support, in other words, the theory of the "European miracle." Hall uses it in precisely this way.

Hall's discussion of "Islam" has a fairly simple argument structure. The first section describes "classical Islam" both as a society and as a religion. We are treated to a detailed description of the mainly Arabic world, and the classical religion, for the period down to A.D. 1,000 Missing from this is any discussion of the earlier history of peoples who became Islamic. It is as though the social category "Islam" emerged, entire, at the time the religion was born. (The stereotype, more precisely, is that of nomads spreading outward from mysterious Arabia.) Thus there is no reference to ancient Egypt, Persia, and so on, as cultural roots. By neglecting pre-Islamic civilization, and then by neglecting later history after the classical period, Hall obtains a model of a nomadic type of society with its own distinct religion coming to dominate a vast region and thereafter giving the region its permanent character. This model also permits Hall to neglect, in most respects, the special civilizational qualities of all of today's Islamic peoples—Pakistanis, Iranians, Indonesians, Nigerians, and the rest, leaving only "Islam and Pastoralism."

Hall finds in classical Islam "a distrust for the exercise of political power." This reflected the fact that its originators were Arab tribesmen who "did not 'feel at home' in their conquered lands." They were united by their religion but were tribal, wandering people, and so "government in Islam tended to be highly unstable" (p. 29). Conceding that the later Ottoman Empire was an exception in this regard, Hall considers this instability to have been a general feature of "Islam" throughout its history. Religion, not the state, tended to be the source of legitimation in society (p. 89). In fact, says Hall, there really *was* no society:

> And what *was* society? It was a large cultural area within which polities of various sizes came and went. . . . [It] was an area held together by an ideology. (p. 89)

Thus we have an image of mute, irrelevant cultures, without real society, without real government, historyless, ruled by alien tribesmen and an imported religion.

Having painted the classical picture, Hall now turns—as is now almost obligatory in Eurocentric treatments of the Middle East—to Ibn Khaldun. Khaldun was indeed one of the greatest social theorists of all time, but his high reputation among Eurocentric scholars such as Hall rests on the fact that he described an Islamic society, the urban society of the Maghreb in the fourteenth century, which was in a state of decline; that he analyzed that decline; and that he predicted, in essence, that the decline would continue. So he gives the Eurocentric historian some useful arguments, legitimated by the quality of Khaldun's reasoning and the important fact that he himself was a Muslim. Khaldun's analysis was indeed valid for the unique case of small trading cities on the North African coast that were in decline because of the shift of economic and political power, during that period, toward the east and the rising power of Iberian Christian states to the north. But it cannot be generalized to other Muslim societies of that time or later. Khaldun states that much of

the cause of the problem lies in the effect on these cities of nearby pastoral tribesmen, who contribute to instability and rather dishearten the urban elite, the overall effect being general stagnation and decline of civili-
zation. For Hall this argues, first, that nomadic tribesmen are a general
and permanent cause of insecurity and various other problems *throughout*
Islam, and, second, that stagnation, decline, is a general characteristic *of*
Islam—and he believes that Khaldun proved both arguments to be true.
This generalization is completely unfounded, both as a deduction from
Khaldun and as a characterization of the Muslim societies of the Middle
Ages and the early-modern period. In short, having described the so-
called pastoral nature of classical Islam and projected it forward in time,
Hall next adds the special description of the medieval Islamic city sur-
rounded and harassed by nomadic tribesmen, thus projecting forward
into the Middle Ages (and later) the idea of stagnation, instability, and
inability to modernize.

This is given a special twist when Hall talks about the Islamic city
(still using Khaldun as his springboard). Hall claims that this city had no
autonomy, was in perpetual semichaos because it was ruled by aliens and
was, therefore, unable to evolve an urban economy like that of Europe. In
fact, his descriptions of the so-called Islamic city are as valid (or invalid)
for medieval European cities as they are for Middle Eastern ones: few of
the former were free of political rule by the lords, and the development of
capitalism in and around these European cities took place in the early-
modern period, when the important Middle Eastern cities—not the declining trading ports of North Africa—were bustling with protocapitalist
activities. In the late fifteenth century Cairo was one of the most important industrial cities in the world. Other Islamic cities, some of them
city-states and thus truly autonomous, were intensely active in world
trade, more so than European cities with the possible exception of Genoa
and Venice. The development of a bourgeoisie or merchant class, with
their allies, the commerce-minded landlords, and of the class relations
characteristic of protocapitalism at the end of the Middle Ages, was certainly as far along in these Islamic cities as it was in Europe. Hall claims
that there was no sense of "liberty" in the Islamic cities, in contrast to the
European ones, and this is nonsense. He goes so far as to claim that the
absence of autonomy in the Islamic city tended to block the development of technology and science (although technology in most spheres
was at least as high and probably higher in the great Middle Eastern cities
at the end of the Middle Ages than in the greatest European cities); going further still, he claims that this blockage derived in part from the very
nature of Islamic doctrine, which, irrationally, downplayed "natural law"

because it claimed, he says, that God interferes in the world (p. 101). Did not Christianity make the same claim?[17]

Hall sums up the "blocking" effect of Islam and Islamic society on economic development in mainly the following propositions, none of them valid. First, he claims that the physical environment had something to do with it. Nothing like the "northern European clay soils" were to be found in the Islamic world (p. 99). These northern European soils, in fact, were not even unusually productive, and across the Middle East and much of Asia there were some zones with soils much higher in agricultural potential than the typical soils of northern Europe. Second, he associates the supposed instability of Islamic politics with the *iqta* and *wafd* landholding systems, claiming that the pattern of landownership in Islam was too unstable to permit landowners to invest in agricultural improvements. A generalization as large as this for all of Islam could not in any case be taken seriously. In fact, however, medieval agriculture in many parts of the Muslim region was highly developed and highly commercialized; instability of land tenure was probably less than it was in chaotic feudal Europe, with its wars, its complex feudal landholding patterns, and the like; and, in any case, the peasants, who did most of the agricultural innovating in those times, tended (though not always) to be somewhat immune to the political instability of most regions in all continents. Third, Hall makes the argument we have just discussed about the city and its lack of autonomy, of stability, and the like.

The basic causality, in all of this, is partly a matter of religion, reminding us of Weber, and partly a matter of political instability ("tribalism" and so on), a lack of the kind of state that, says Hall, is needed to bring about development.[18] Hall finds irrationality in Islamic society, as he does in Indian and Chinese society. And he finds Oriental despotism, as well: a "predatory," unstable state, given to irrational warfare ("In Muslim society, wars always remained the greatest potential source of profit," p. 102.) Therefore no possibility of progress toward modernity.

THE RISE OF CHRISTIAN EUROPE

So much for China, India, and Islam. Hall now turns to miraculous Europe. He throws in a large number of different causes for the "European miracle," although the most stress is laid on Christianity—in a familiar Weberian argument, slightly modified—and on the European state, this also seen as a product mainly of Christianity. Hall, however, is eclectic, and the argument cannot be considered simply religious determinism. In fact, he seems to want to include as many reasons as he can think of for the unique rise of Europe. He begins with Europe's natural environment.

Europe was environmentally favored. This continent

> is a divided area with several small cores, the majority of which have deep and productive clay soils fed by rainfall. There was no need for irrigation. It is quite likely that this encouraged, or at least allowed for, a decentred agricultural civilisation based on individual initiative. (p. 111)

We disposed of these empty environmentalistic arguments in our discussion of Jones's book *The European Miracle*, which clearly is Hall's source.[19] As we saw, the idea of "ecological cores" has no causal significance as to the uniqueness of Europe's development, and the pattern of relatively fertile cores surrounded by hilly or swampy peripheries is by no means unique to Europe. Nor is Europe uniquely "divided." (Notice, for instance, Southeast Asia, with its peninsulas, its archipelagos, its capes and bays.) I also commented in Chapter 5 on the myth that Europe's "clay soils" are somehow especially productive. These, for the most part, are gleys and podzols, which tend to be waterlogged and acidic, and, although they are perhaps on the average moderately productive under careful management, they are not superior, on average, to the agricultural soils of many parts of Africa and Asia. As to irrigation, again we have the classic myth of "Oriental despotism." Hall rejects Wittfogel's famous argument that irrigation leads to strong states, but he fully accepts the basic (and truly classic) thesis, that the "need" to irrigate forces societies into the yoke of despotism because, supposedly, irrigation requires a command structure to maintain the waterworks. But there is no "need" to irrigate. Societies sometimes determine that the elaboration of irrigation systems leads to greater production, prosperity, and social progress (at least for the ruling elites). Most early irrigation systems were small in scale, and probably were controlled by villages and small intervillage councils, not by overarching state structures. The idea that the technology of irrigation somehow calls into being a despotic government is a false causal argument grounded only in the fact that early civilizations had despotic governments—they were not what we would think of as democratic— and *some* of these civilizations *did* develop elaborate irrigation systems as they evolved socially. And there is no truth to the notion that Europe's soils somehow led to "a decentred agricultural civilisation based on individual initiative." Europe had no environmental advantage over Africa and Asia.[20]

Hall gives a lot of credit to the Roman empire as a fountainhead for the rise of Europe, in this respect diverging from the arguments of most of the other historians whom we discuss in this book (most notably Jones, Mann, and Landes, for whom Rome was another one of those "empires"). We are told about a cornucopia of civilizational innovations emanating in part from the empire, in part from the early Church. First, Hall contrasts Rome favorably with ancient China. It was more cosmopolitan and pluralistic (not true). It brought civilization to barbarian lands whereas China did not (again not true). It had more of a tradition of law, more of rebellion, more of tolerance, than China (untrue). It had a greater economic potential, grounded in literacy and the use of coinage (untrue: both present in China). But the main virtue of Rome was its role as cradle of the Christian Church, and Hall explains to us how the early church brought progress to Europe in ways that other churches, and other institutions, did not do elsewhere—again a comparative judgment based on ignorance of non-Europe. The egalitarian nature of the very early church is not of course in doubt, but Hall sees this as the source of a democratic tradition in later European history (ignoring both the socially conservative role of the medieval church and the equally egalitarian nature of Islam during that period). Hall contrasts the willingness of Christian missionaries to bring civilization to the barbarians and the supposed unwillingness of Chinese to do the same, this in the face of our knowledge of the proselytizing Buddhists and later Muslims in and around China. Summing all of this up, Hall says, "The church wore the mantle of Rome. It *was* civilization" (p. 120). True, but not as an argument for a European miracle.

Hall next focuses on what he considers to have been a technological revolution in northern Europe (and nowhere else) during the early Middle Ages. His point here is that this occurred because of the characteristic dynamism of Europe's economy and was associated closely with the rise of an autonomous market—a matter that we will discuss in a moment. The supposed technological revolution consists of basically the same menu of innovations that were listed for us by Lynn White, Jr., and Eric Jones. The "water mill" was known to the Romans but was really put to proper use, according to Hall, in medieval northern Europe, where it proved that "there was *considerable* investment at the local level" (p. 121). (It was known and used in other parts of the Eastern Hemisphere.) The heavy plow also, says Hall, was known to the Romans but really put to use later in northern Europe, with all of the marvelous effects noted by White and Jones. (Recall, from our earlier discussion of the matter in Chapter 3, that the heavy plow was used a thousand years earlier in India, and that the cause-and-effect argument from plow to society is untenable.) All told, the technological revolution demonstrated that "this society showed considerable skill at invention . . . and even more at adopting and adapting inventions which it borrowed from Islam and China" (p. 122). The implication here is that other societies were less inventive, and moreover were less willing to borrow from their neighbors—again the Weberian claim for European rationality. That the technological process was causally crucial was demonstrated, says Hall, by the rise of population. And indeed technological progress did take place and did lead to a rise of population, but it did so in the same period elsewhere. Hall, however, sees population rise elsewhere as a Malthusian affliction. Population growth was progress in Europe but disaster elsewhere.

Hall now introduces the traditional argument that feudal landholding patterns led to unique economic development in Europe. He lays some emphasis on what he calls the security of land tenure patterns, claiming, quite falsely, that political stability was greater in medieval Europe than in Asia and so landlords were able to hold on to their land, to invest, and so on. From here he turns to the Weberian—actually much older—argument that feudal landholdings were closer to genuine private property than were the service tenures supposedly characteristic of the Asian empires. We have already seen (in Chapter 2) that this contrast is a false one: Asian land tenure was as close to private property as was European in this period; service tenures were characteristic of feudal Europe and both there and in Asia tended to become hereditary properties.[21] To this picture Hall adds, for Europe, a sturdy yeoman peasantry, "freemen . . . with some of their own land," playing a key role in economic progress (p. 128). This is a simple telescoping of history. Medieval peasants were not, in general, freeholders or "yeomen," but rather were serfs and tenants carrying heavy burdens of rent paid in labor, produce, or cash. Progressive they indeed were, but Hall's conception is that of a truly modern, individualistic, entrepreneurial, landowning, capital-investing, yeoman farmer as *the* characteristic small farmer of the Middle Ages. None of this was true. And the picture of a somehow ebullient, rapidly progressing economy deep in the Middle Ages is not really valid. There was some progress, of course, but there was comparable progress outside of Europe as well.

Hall next asserts that the European family was unique, and contributed to Europe's unique economic rise. He ascribes to the European family a unique ability to avoid the Malthusian trap of overpopulation. At the root is "the relative continence of the European family" (p. 131). He is telling us that Europeans practiced sexual continence and others did

not—hence, telling us (as Jones did) that non-Europeans are either irrational or incontinent in their unwillingness or inability to restrain their
sexual urges and so limit their number of children. Beyond this, Hall argues that the European family was unique in two respects: it was small
(nuclear, not extended), and it was relatively unimportant as an institution within the society as a whole. Whether or not he agrees with Jones
that the origin of Europe's unique family system goes back to prehistoric
times is unclear. He echoes Jones in asserting (falsely) that in Europe
"improvements in output were not eaten up by a massive growth in population," as they were, he says, in China (p. 131). But most curious of all is
Hall's assertion that the nuclear family somehow reflected a *weak* kinship
system, and this in turn led to a *strengthened* European state. In Europe,
he says, the ordinary people did not have a strong kinship system with
which to defend themselves; this "made the European peasantry that
much better fodder for state formation" (p. 33). This is another illogical
argument, and one grounded in a false belief, namely, the idea that the
European family was in fact unique. As we discussed previously, it was
neither unique nor perhaps even unusual.[22]

If Christianity is seen by Hall as the main underlying force in the European miracle, the state is seen as the foreground institution. We have
already observed how he dismisses the forms of state that were found in
China, India, and Islam. These regions either had rather superficial polities or, worse, they had empires. (Recall his vehement comment: "Imagine what European history might have been like . . . had any empire"
been imposed.) The European state, says Hall, was neither too weak nor
too strong, but was rather (as for Goldilocks) just right. It was an "organic
state." He devotes some space to defining this notion, but in the end it
remains nothing more than a value judgment. The European state of the
Middle Ages was "organic" because, Hall thinks, it did the right things
for the society: it gave peace, gave services, and the like. But it did not do
much of that. These qualities are truly found in the modern state, from,
perhaps, the seventeenth century onward. Like all of the other Eurocentric historians whom we have been discussing, Hall wants to push back
into the Middle Ages many of the positive virtues of European society
that emerged *after* the rise of Europe, after Europe had well begun its economic modernization. In this way the false argument is made that, somehow, the seeds of modernization were present very early in Europe and
not elsewhere. But probably there were no well-integrated states in Europe during the time that Hall is discussing. These matters are of course
relative, but whatever degree of political integration there may have
been in countries like, say, Britain and France, there were certainly com-

parable levels of integration in other continents. Nor can we accept Hall's argument that empires were somehow not well integrated, that they were despotic but nonetheless weak, and so less "organic" than the European state. He produces no evidence for this statement; indeed, his discussion of the matter, as we saw, is based on lack of knowledge of Asian history.

Not only was the European state unique, says Hall, but so, too, was the system of states. Here he repeats Jones's argument about the marvelous character of Europe's supposedly unique interstate system in the Middle Ages, an argument that, as we saw in Chapter 5, is untenable because there really was no system of states until early-modern times and until considerable modernization had occurred.

Many other factors are invoked by Hall to help in explaining the "European miracle," but these require very little discussion. Hall thinks that "rational science" was a special product of the Judeo-Christian tradition. "Rational science . . . was to some extent blocked in other world civilizations." Not in Europe. Here, he suggests, the peculiarly Greek idea of natural law became married to the peculiarly Judeo-Christian idea that God does not "habitually interfere with the rules of nature" (p. 133). As we saw in our discussion of other historians, mainly Weber and Jones, this is just another one of those prejudices against non-European cultures. Scientific thought was characteristic of all the major civilizations.[23] Modern science, in Europe, emerges well after economic modernization has begun.

It remains only to mention some arguments which Hall makes concerning the ways Christianity is supposed to have aided Europe's medieval economic development. (That it did just that is not at all in dispute. Hall is claiming, rather, that the result was a miraculous, uniquely European, economic development, and this we dispute.) Some of the supposed roles played by the religion and the church have been discussed already: for instance, effects on science and on the civilizing of barbarians. Hall adds further roles, some of which it did indeed play in Europe's development, some of which it did not. But Hall wants to make one very strong point about the importance of Christianity in relation to the European state:

Christianity provided the best shell for the emergence of states. (p. 135)

This supposedly is in contrast to the other great religions in other regions. The only real argument given for this quite strange assertion is a mention of the way that Christianity legitimized rulers, crowned the kings, and the like. Did not all religions do roughly the same? Christianity, says Hall, "kept Europe together" after the fall of Rome (p. 123). True, but other religions played the same culturally cohesive role in other societies. But Christianity

> differs from Islam . . . [and] Hinduism, since it did not "block" politics, and so did not encourage a climate of instability which limited the autonomy of market relationships. (p. 143)

This is nonsense.

Hall's theory can be summed up quite simply as follows. Capitalism naturally tended to develop in Europe. It was natural in exactly the sense that Adam Smith noted long ago. Europe's environment, Europe's inhabitants (with their "continence," rationality, and so on.), Europe's political institutions and Europe's religious institutions, all played distinctive roles, permitting the natural evolution of a capitalist economy to take place. The people, institutions, and environment in non-Europe "blocked" this development. That is why Europe rose and other societies did not.

Hall's argument is basically a synthesis of the theories of Weber, White, and Jones, making use also of ideas taken from Mann and Brenner. It is a sociological stewpot, lightly seasoned with politics.

NOTES

1. Hall, *Powers and Liberties: The Causes and Consequences of the Rise of the West* (1985). Hall, "States and Societies: The Miracle in Comparative Perspective" (1988). Also see Hall and Ikenberry, *The State* (1989).

2. Hall, "States and Societies," pp. 24, 38; Hall, *Powers and Liberties*, pp. 141–144 and throughout.

3. Hucker comments that "Ming government apparently put a light burden on ordinary Chinese. . . . Considering how it maintained its power and sustained its subjects both morally and materially, the Ming government probably deserves to be reckoned, on balance, the most successful major government in the world in its time"; in "Ming Government" (1998), p. 105. Also see Pomeranz, *The Making of a Hinterland: State, Society, and Economy in Inland North China 1853–1937* (1993); Rowe, *Hankow: Commerce and Society in a Chinese City, 1769–1889* (1984); Marks, *Tigers, Rice, Silk, and Silt: Environment and Economy in Late Imperial South China* (1998); Subramanian, "India's International Economy, 1500–1800" (1999).

4. Purcell, *The Chinese in Southeast Asia* (1951). This matter is discussed further in Chapter 8.

5. The statement is made without citing authority. Apparently he relies here on Jones's book *The European Miracle* (copiously cited by Hall), in which this statement is

made as a result of misreading Filesi, *China and Africa in the Middle Ages* (1972). See Brook, "Communications and Commerce" (1998): "Maritime trade flourished in the mid-Ming in spite of government bans" (p. 696). Hall also asserts, wrongly, about the Chinese economy during this period that the Ming government abandoned coinage and created "a purely natural economy" (p. 50). See Von Glahn, *Fountain of Fortune: Money and Monetary Policy in China, 1000–1700* (1996).

6. Frank, in *ReORIENT* (1998), cites various sources that confirm this fact. Also see Pomeranz, "De Long on David Landes" (1998).

7. Another strange statement about Chinese culture: the fact that China used a nonalphabetic written language somehow led to the "extreme cohesion of the scholar-gentry class." Even more ignorantly: China, unlike Rome, had "the morpheme . . . to bind the elite firmly together"; in Hall, *Powers and Liberties*, p. 113. All languages have morphemes.

8. See, for instance, Baechler, "The Origins of Modernity: Caste and Feudality (India, Europe and Japan)" (1988), pp. 39–66.

9. Hall's tendency to use stereotypes and traditional Eurocentric notions about India must be connected to the fact that he relies on old and mainly European sources, as indicated by his bibliography. Citations in his book are to works on the average about thirty years old, and very few of them are to Indian sources. This rather throws into question his habit of pronouncing upon the poor quality of historiographic sources for India: "We lack written records of early Indian history, and the historiography of India is likely to remain the weakest of all the world civilisations. A part of the power of the Brahman in Indian life has been based on the capacity to memorise and recite rather than to refer to written documents" (p. 58). "[The] relative absence of documents in India means that the account offered must necessarily be uncertain" (p. 78). None of this is true. The deficiency is Hall's, not India's.

10. On the same page he writes of the "irrelevance of political power in India," and says of Indian politics, there "was much sound and fury but it really did signify nothing."

11. Subrahmanyam, *Merchants, Markets, and the State in Early Modern India* (1990); Subramanian, "India's International Economy: 1500–1800" (1999).

12. In Hall and Ikenberry, *The State*, p. 80. On this matter see Habib, "Merchant Communities in Precolonial India" (1990); Subrahmanyam, *Merchants, Markets, and the State*.

13. See, for example, Dirks, *The Hollow Crown: Ethnohistory of an Indian Kingdom* (1987). It should be noted also that Muslims, not Hindus, dominated the coastal trading economy, not to mention the Mughal Empire.

14. Perlin, *The Invisible City* (1993). See Habib, "Merchant Communities in Precolonial India" (1990); Subrahmanyam, *Merchants, Markets, and the State*.

15. Weber is cited abundantly throughout *Powers and Liberties*.

16. Interestingly, Hall contradicts himself about the relationship between Christianity and politics. On p. 28 Christianity was "irredeemably involved in politics." But on p. 29, "Christianity . . . said that the purpose of religion was purely spiritual, and . . . power relations did not matter and could be left to proceed on their own course." He finds it useful to stress Christianity's political involvement for his attack on India and its political noninvolvement for his attack on Islam.

17. On technological development in Islamic regions, see Watson, *Agricultural Innovation in the Early Islamic World: The Diffusion of Crops and Farming Techniques, 700–1100* (1983); Al Hassan and Hill, *Islamic Technology* (1986).

18. Hall makes a special argument concerning the Ottoman state, which, Hall ad-

mits, does not follow the pattern of instability painted by him for Islam as a whole. He merely announces that this state had the problems of "empire," as he had argued for China; beyond this, he quotes E. L. Jones to the effect that the Ottoman state was in general nasty.

19. Hall does not cite Jones in support of this precise passage but does so throughout his book and article.

20. Other environmentalistic arguments are given as well. For instance, echoing Jones, Hall states that Europe suffered less from natural disasters than did Asian societies. We dealt with this myth in Chapter 4. Also see Volume 1, Chapter 2.

21. See Kumar, "Private Property in Asia" (1985).

22. See the discussion of the family in Volume 1, pp. 128–135, 149–151.

23. See, on this topic, Goody, *The East in the West* (1996); Needham, *Science and Civilization in China* (1954–); Sivin and Nakayama, *Chinese Science* (1973).

CHAPTER 8

Jared Diamond:
Euro-Environmentalism

ENVIRONMENTAL DETERMINISM

The theory of environmental determinism, or environmentalism, has played a crucial role in Eurocentric distortions of history. This theory does not simply assert the obvious fact that the natural environment is part of, plays a role in, every human act. Environmentalism is the practice of falsely claiming that the natural environment explains some fact of human life when the real causes, the important causes, are cultural, not environmental. Our concern here is with environmentalistic theories of history and more especially with theories that falsely claim that the natural environment of Europe is superior to that of other parts of the world, superior in the sense that it has led Europeans to progress further or faster than other peoples, who supposedly occupy inferior environments. Let us call this sort of argument "Euro-environmentalism."

Euro-environmentalism is one of three basic or foundational theories that have been used in the past century to explain the superiority or priority of Europeans in history. The others (as we noted in Chapter 1) are biological racism, the theory that Europeans inherit their superiority through their genes, and culturalism, the theory that European culture has been, for whatever ultimate reason, superior to all other cultures since time immemorial. These three theories often are used in combination with one another. In the days when Christian Europeans tended to believe that their historical progress is uniquely guided by a Christian god, it seemed reasonable to believe that God arranges for Europe to have a superior environment as well as superior heredity and a superior culture.

Environmental determinism was not seen as materialistic and atheistic, but as one of God's instruments.

After about the middle of the nineteenth century, historical theorizing no longer tended to invoke the Deity, but most Eurocentric explanations still consisted of a combination of the three foundation theories, race, culture, and environment. Race dropped out of most explanatory arguments after World War II, owing in part to the association of racism with Nazism. Today almost all Eurocentric theories of history argue that environment and culture have jointly worked to produce Europe's superiority or priority in history. Some formulations, including those of Eric L. Jones (Chapter 5) and David Landes (Chapter 9), use both environment and culture in more or less equal parts. Others, such as those of Max Weber (Chapter 2), Michael Mann (Chapter 6), and John A. Hall (Chapter 7), tend to stress cultural explanations but throw in a good bit of Euro-environmentalism as well. Theories that explain Europe's supposed historical superiority rather *strictly* in environmentalistic terms, theories that are reasonably *pure* examples of environmental determinism, used to be quite popular, but this is no longer the case. One such theory, however, was put forward by Jared Diamond, in his much discussed Pulitzer Prize-winning 1997 book, *Guns, Germs, and Steel: The Fates of Human Societies*.[1] This chapter is a critique of Diamond's theory and of Euro-environmentalism in general.

"Environment molds history," says Diamond flatly and without qualification (p. 352).[2] Everything important that has happened to humans since the Paleolithic Age is due to environmental influences. More precisely: all of the important differences between human societies, all of the differences that led some societies to prosper and progress and others to fail, are due to the nature of each society's local environment and its geographical location. History as a whole reflects these environmental differences and forces. Culture plays a much smaller role: the environment explains all of the main tendencies of history, with cultural factors affecting only the minor details. Diamond proceeds systematically through the main phases of history in all parts of the world and tries to show, with detailed arguments, how each phase, in each region, is explainable largely by environmental forces. The final outcome of these environmentally caused processes is the rise and dominance of Europe.

The essential argument is very clear and simple. Almost all of history after the Ice Ages happened in the temperate midlatitudes of Eurasia (the landmass containing Europe and Asia). The natural environment of this large region is better for human progress than are the tropical environments of the world, and the other temperate regions—South Africa, Australia, and midlatitude North and South America—could not be central for human progress because they are much smaller than Eurasia and are isolated from it. Although many civilizations arose and flourished in temperate Eurasia, only two were ultimately crucial, because of their especially favorable environments: China and Europe. Finally, some five hundred years ago China's environment proved itself to be inferior to Europe's in several crucial ways. Therefore Europe in the end was triumphant.

Guns, Germs, and Steel is written for a popular readership, and Jared Diamond, who is a natural scientist (a biologist), structures the book as a sequence of chatty lectures, each of which is designed to show his readers that some important problem in human history is easily solved when we look at it *scientifically.* And doing so, we will find that the explanation is quite simple, and it starts with the natural environment. To convince nonscientists that he is right in these matters, Diamond employs a number of devices that make his assertions seem impressively scientific. I will discuss these devices as we work our way through his narrative, but one of them appears at the very outset of the argument: a pseudoexperiment.

"A NATURAL EXPERIMENT"

"A Natural Experiment of History" is the title of the first substantive chapter of *Guns, Germs, and Steel.* Diamond will compare New Zealand's Maoris with a small community of people of the same cultural origin, the Moriori, who, several centuries ago, settled the Chatham Islands, and he will produce a natural experiment that will show that the difference between the two environments explains the difference in their subsequent history.

> Moriori and Maori history constitutes a brief, small-scale natural experiment that tests how environments affect human societies. Before you read a whole book examining environmental effects on a very large scale—effects on human societies around the world for the last 13,000 years—you might reasonably want assurance, from smaller tests, that such effects really are significant. If you were a laboratory scientist studying rats, you might perform such a test by taking one rat colony, distributing groups of those ancestral rats among many cages with differing environments, and coming back many rat generations later to see what happened. Of course, such purposeful experiments cannot be carried out on human societies. Instead, scientists must look for "natural experiments," in which something similar befell humans in the past. Such an experiment unfolded during the settlement of Polynesia. (pp. 54–55)

The climate of the northern part of New Zealand is relatively warm, and the Maoris, who arrived as agriculturists, could and did practice agriculture in this region. The Chathams are a group of tiny islands 300 miles to the east of New Zealand; they are small, cold pieces of land, unsuitable for agriculture, and as a consequence the Moriori have "reverted" (p. 54) to hunting and gathering. Diamond asserts that the two societies have been isolated from each other for centuries, so they can be compared as two experimental conditions. Europeans discovered the Chathams in the nineteenth century, and, hearing the news, some Maoris sent out an expedition to conquer the islands and enslave the inhabitants. They succeeded in this endeavor. This demonstrates, according to Diamond, that peoples who live in an environment favoring agriculture, and who practice the higher art of agriculture, will have larger populations, higher technology, more power, and greater all-around success than hunter-gatherers.

The word "experiment" has no place in this discussion, for three reasons. First: this is not an experimental treatment; just a comparison, puffed up to sound more scientific than it really is—a device often described as "scientism." Second: the comparison cannot be reduced to the action of two, or even a few, variables, here called "environment" and "culture." And third: the entire scenario can be described in a way that is altogether unsurprising. It appears that Morioris migrated from the southern part of New Zealand's South Island, a region that is too cold to sustain crop agriculture. In fact, I would expect that Maoris in that region used to employ subsistence activities much like those of the Chatham Island Morioris: hunting, gathering, fishing, shellfishing, sealing.[3] (I wonder, also, whether Maori communities in the cold southern part of South Island suffered the same aggression at the hands of some warlike Maoris from the agricultural northland as Moriari did at the hands of northland Maoris.) Moreover, the Chatham Islands are not "subantarctic," as Diamond describes them (p. 58): they lie about 45° south latitude. They used to sustain dense broadleaf forest, and the local resources were so rich that Morioris made quite a good living harvesting them. In a word, it is hardly the case that Maori and Moriori occupy, as it were, two separate petri dishes to be used as contrasting experimental conditions. If Diamond merely asserted the comparison between the size and strength of populations practicing hunting–gathering and those practicing agriculture, he would present us with an obvious fact. Actually, he belabors this obvious fact throughout the book, and uses the comparison between hunting-gathering and agriculture to explain (or so he claims) an incredible variety of historical and geographical facts. In this introductory chapter, he sets the tone of the argument: the treatment will be experimental; scientific. How can you argue with science?

AGRICULTURE

Diamond distinguishes between the "ultimate factors" that explain "the broadest patterns of history" and the "proximate factors," which are effects of the "ultimate factors" and explain short-term and local historical processes (p. 87). The "ultimate" factors are environmental, not cultural. The most important of these "ultimate" factors are the natural conditions that led to the rise of food production. Those world regions that became agricultural very early gained a permanent advantage in history. Regions that became agricultural much later were at a permanent disadvantage, but those that never acquired agriculture on their own were, for that reason, totally out of the stream of historical development. The "ultimate" causes led, in much later times, to regional variations in technology, political organization, and health; these, then, were the "proximate" causes of modern history. More than half of *Guns, Germs, and Steel* is devoted to elucidating the "ultimate" causes, explaining why differing environments led to differing rates in the acquisition of agriculture, and explaining how the resulting differences produced several thousand years of human history. I will try to show that this argument is very flawed. But notice first the crude scientism in this formulation: thousands of years of cultural evolution in a region cannot overcome the effects that the natural environment supposedly had on the region back in the Neolithic period. Culture is a feeble force in history.

The "ultimate" causes are three primordial environmental facts: the shapes of the continents, the distribution of domesticable wild plants and animals, and the geographical barriers inhibiting the diffusion of domesticates. The first and most basic cause is the shape of the continents: their "axes." A continental landmass with an "east–west axis" supposedly is more favorable for the rise of agriculture than a continent with a "north–south axis." Diamond divides the inhabited world into three continents (he uses the word "continent" rather broadly[4]): Eurasia, Africa, and the Americas.[5] Eurasia has an east–west axis; the other two have north–south axes. This has had "enormous, sometimes tragic consequences" for human history (p. 176).

Actually, Eurasia is almost as tall (north–south) as it is wide (east–west), a matter of about 5,000 miles as against 7,000 miles, and if North America were treated as a continent its north–south and east–west dimensions would be about equal. It becomes evident as we read the chapter in *Guns, Germs, and Steel* titled "Spacious Skies and Tilted Axes" that Diamond is not talking about axes at all; he is making a rather subtle argument about the climatic advantages that (in his view) midlatitude regions have over tropical regions. The world's largest continuous zone of "temperate" climates, those that supposedly are neither too hot nor too cold, neither tropical nor subarctic, lies in a belt stretching across Eurasia from Europe in the west to Japan in the east. Rather persistently neglecting the fact that much of this zone is inhospitable desert and high mountains, Diamond describes this east-west-trending midlatitude zone of Eurasia as the world region that possessed the very best environment for the invention and development of agriculture and, consequently, for historical dynamism, because—and this is hardly controversial—an agricultural mode of subsistence allows for settled communities, dense populations, and various consequences thereof.

Why would one expect the origins and early development of agriculture to take place in the midlatitude belt of Eurasia? Diamond posits a number of environmental reasons, which I will examine in a moment. First, however, comes the question where agriculture actually did originate. Diamond notes, correctly, that there are thought to have been several apparently independent centers of origin, only two of which lie in the temperate belt of Eurasia. These two are the Near East (the "Fertile Crescent") and China. Diamond needs to show, for his central argument about environmental causes in history, that these two midlatitude Eurasian centers were earlier and more important than tropical centers (New Guinea, Ethiopia, West Africa, Mesoamerica, the Andes, Southeast Asia, India, the Amazon region). And he needs to show, further, that the Fertile Crescent was the earliest and most important center because this region's environment led, by diffusion westward, to the rise of Western civilization. Indeed, at various places in *Guns, Germs, and Steel* the traditional Eurocentric message is conveyed that the Fertile Crescent and Mediterranean Europe are a single historical region; that history naturally moved westward from the one to the other.

Now the question of where, when, and how often agricultural revolutions took place is by no means settled. Probably the majority of specialists think it likely that the Fertile Crescent was the earliest such center, but all are aware that there are alternative candidates with good credentials. Domestication has been traced back to about 8500 B.C. in the Fertile Crescent; the earliest dates thus far obtained for New Guinea and China are not very much younger: 7500 B.C. for New Guinea and 7000 B.C. for China. But consider the fact that archaeologists have been digging in the Near East for a couple of centuries; vastly more data have been obtained for this region than for any other; and the proposition that agriculture originated there perhaps 10,000 or 11,000 years ago has been accepted for some time. By contrast, thirty years ago Chinese agriculture was thought to be perhaps only five or six thousand years old, and nobody thought that New Guinea agriculture had any antiquity whatever. The point is that archaeology in these and many other regions is very incomplete, and one must consider it very likely that much earlier dates for the agricultural revolution will soon be accepted for many regions other than the Near East. As to the origins of agriculture in the humid tropics, it is very difficult to determine archaeologically how early the agricultural revolution or revolutions occurred, mainly for two reasons: first, plant remains and other organic residues are much less well preserved in this warm climate; and second, there is the intriguing fact that a vast plain in Southeast Asia, the shallow portion of the Sunda Shelf, was dry land until perhaps 7,000 years ago,[6] and may well have been one of the primary centers of early agriculture before it was inundated by rising sea levels.

Diamond essentially ignores these uncertainties about time and place. For him, agriculture arrived first in the Fertile Crescent; China, perhaps independently, had its agricultural revolution somewhat later; and all other regions are later still. This posture allows him to develop some of his most basic environmentalistic arguments. The Fertile Crescent, he argues, developed agriculture first mainly for two reasons. It has a Mediterranean climate (hot, dry summers and mild winters with rainfall concentrated in the winter months). And many of the important domesticable wild cereal grasses, in particular the wild ancestors of wheat and barley, are native to this region. Just why a Mediterranean climate is especially favorable for agricultural origins is not made clear. Apparently Diamond thinks that the winter-wet Mediterranean regime favored cereals with large seeds, but other staple cereals, found wild in other climates, and notably maize, rice, and some sorghums, also have large seeds, while many small-seeded cereals, such as many varieties of millet, are staples in other regions, again with other types of climate. (Noncereal staple crops, like yams and potatoes, are dismissed by Diamond as unimportant, for reasons that we will go into in a moment.) Diamond's emphasis on Mediterranean climates has a teleological ring to it: if agriculture originated in this type of climate, then history would naturally move westward, not eastward and southward, because southern Europe also has a Mediterranean climate, whereas the other temperate regions of Eurasia have summer-wet climates.

The argument about Eurasia's "east–west axis," as we saw, is really an argument for the historical primacy of the midlatitude environments of Eurasia. Although the Fertile Crescent is privileged, there is a somewhat broader argument that encompasses the whole region, from Europe at one end to China at the other. This argument is a two-part claim that, first, the domestication of cereal crops was more important for history than that of other staples (notably yams, potatoes, taro, manioc, sweet potatoes, and bananas), and second, the cereal crops that were domesticated in midlatitude Eurasia (especially wheat in the Near East and millet in China) were more important than other cereals (notably rice, sorghum, and maize) that were domesticated elsewhere. These other staple crops were domesticated in tropical or subtropical regions: rice somewhere in mainland Southeast Asia or adjoining regions of south China and India; sorghum in the Sudanic region of sub-Saharan Africa; maize in Mesomerica. As to noncereal staples: yams were domesticated in West Africa (and less important species of yams elsewhere in the tropics); potatoes probably in the Andes; manioc in the Amazon region; sweet potatoes in tropical America; taro and bananas in tropical Southeast Asia.

But Diamond argues that wheat and millet were much more important for history than these other crops, and this goes far toward explaining the rise, respectively, of the West and China. He argues unpersuasively that rice and maize are poorer in protein content than wheat; but the difference in fact is rather small, mostly a matter of moisture content. He makes a strange argument about maize: since early domesticated varieties had small cobs and kernels, it must follow that maize—presumably the oldest New World staple—took much longer to become a fully domesticated crop and did not reach that stage until long after agriculture had arisen elsewhere; this would fit his theory favoring nontropical environments. (Maize was domesticated in tropical Mesoamerica.)

Diamond's argument against the root and tuber staples (potatoes, yams, and so on) dredges up an old and discredited theory: these crops are much higher in starches and lower in proteins than cereals; hence, the people who depended on them in the early days supposedly were not properly nourished and so cultural development in these regions must have been inhibited. In fact, peoples who use, and used, these root and tuber crops eat rather great quantities of the foods (which have very high moisture content) to obtain most of their nutrients, and use other crops, and in some cases also domestic animals, to obtain additional protein.[7] Diamond is wrong to argue that root and tuber staples are inferior to cereals for human needs. But root and tuber crops, along with maize, rice, and sorghum, were the main domesticated staples of the humid tropics and subtropics; therefore, for Diamond, the tropics were of no real significance in the rise of food production.

The last of the three "ultimate factors" that go far toward explaining "the broadest patterns of history" is geographical diffusion. Diamond invokes diffusion in arguments that need it: when he wants to demonstrate that the spread of some domesticate, or some technological trait, or some idea was rapid and consequential. He neglects diffusion when it is convenient to do so: when he wants to emphasize the supposed isolation of some region (like Australia and the Chatham Islands) and the consequences of that isolation. As regards the rise and development of food production, Diamond's central point is that the relative similarity of the environments within Eurasia's temperate belt accounts in large part for the putatively rapid spread of food production throughout this region as contrasted with the rest of the world. He seems not to notice that the agriculturally productive regions within this temperate belt are quite isolated from one another, separated by deserts and high mountains. Moreover, to view this temperate belt as a continuity he must neglect the fact that the central part of the region is not temperate at all: it is tropical India. (North of the Himalayas there were, again, deserts.[8]) Contrary to Diamond's theory, north–south diffusion, which generally meant diffusion between temperate and tropical regions or between temperate regions separated by a zone of humid tropics, was quite as important as east–west diffusion.

Diamond makes the sensible-sounding deductive argument that agriculture will have difficulty diffusing southward and northward between midlatitude Eurasia and the African and Asian tropics because this requires movement between regions that are ecologically very different; hence it must follow that midlatitude crops will tend not to grow very well in humid tropical regions, and vice versa for tropical crops, because they are accustomed to different temperature and rainfall regimes and either need seasonal changes in day length if they are midlatitude domesticates or, conversely, cannot tolerate such changes in day length if they are low-latitude domesticates. This argument is used by Diamond mainly to support two of his theories. One is the theory that tropical regions of the Eastern Hemisphere tended to develop later, and more slowly, than temperate Eurasia. The other is the theory that Southern Hemisphere regions beyond the tropics, notably Australia and the Cape region of South Africa, did not acquire agriculture largely because intervening tropical regions kept them isolated from the Eurasian centers of domestication. The deduction is false; or rather, the effect of the north–south barriers cannot have been very important. The essence of domestication is the changing of crops, by selection and other means, to make them more suitable for the human inhabitants of a region. Always this involves some changes to adapt to different planting conditions. There are, indeed, true ecological limits. But the range of potential adaptation is very wide. Almost any tropical region with distinct dry and wet seasons is potentially suited for most of the major cereals domesticated in temperate Eurasia. Day length is (was) important for some crops, notably wheat, but in most cases adaptations could, and did, remove even this limitation. After all, in early times some kinds of wheat were grown as far south as Ethiopia (not far north of the equator); rice was grown in both tropical and warm midlatitude climates; sorghum, first domesticated in Sudanic West Africa, spread to midlatitude regions of Asia. In the Western Hemisphere, maize was grown by Native Americans all the way from Peru to Canada. Most tropical root and tuber crops had problems spreading to regions that were cold or seasonally dry, but many of these crops, too, adapted quite nicely: think of the potato and sweet potato. Diamond's error here is to treat natural determinants of plant ecology as somehow determinants of human ecology. That is not good science.

Diffusion is also stressed by Diamond as having been a significant factor in early world history, and some of his points are valid. But when, in various arguments, he posits natural environmental barriers as causes of nondiffusion, or of slow diffusion, he makes numerous mistakes. Some of these, as in the matter of north–south crop movements, just discussed, are factual errors about the environment. Other errors are grounded in a serious failure to understand how culture influences diffusion (Blaut, 1987b). Two examples deserve to be mentioned.

"[What] cries out for an explanation is the failure of food production to appear, until modern times, in some ecologically very suitable areas" (p. 93). All of these areas are midlatitude regions that are separated from midlatitude Eurasia by some intervening environment: a tropical belt, a body of water, or the like. Diamond devotes a lot of attention to two such areas: South Africa's Cape of Good Hope and Australia. Why did these two regions remain nonagricultural for so long? In both cases the sought-after explanation is supposed to be a combination of barriers to diffusion and local environmental obstacles. Cultural factors are ignored.

The Cape of Good Hope is a zone of Mediterranean climate (hot summers, winter rain). What "cries out for an explanation" here is the fact that this area, according to Diamond, had the ecological potential to be a productive food-producing region comparable in some ways to the corresponding Mediterranean climate zone of western Eurasia, but remained a region of pastoralism until Europeans arrived. He thinks that this problem of long ago even helps to explain South Africa's modern racial problems. Bantu-speaking agricultural peoples spread southward into South Africa, but, according to Diamond, they stopped precisely at the edge of the Mediterranean climatic region. This region was occupied by the Khoi people who were pastoralists. Why did the Bantu-speakers, who had invaded Khoi lands farther north, not do so in the Cape region and then plant crops there? Why did the Khoi not adopt agriculture themselves? Diamond denies, rightly, that the this had to do with any failure of intellect. The causes, he argues, were matters of environment and diffusion. The crops grown by the Bantu-speakers, here the Xhosa, were tropical, and according to Diamond could not cope with the winter-wet climate of the Cape region. So the Xhosa did not, could not, spread food production to the Cape because of its Mediterranean climate. The Khoi, for their part, did not adopt agriculture because Mediterranean crops that had been domesticated north of tropical Africa could not diffuse through the region of tropical environment and agriculture to the Cape, and because the Cape region did not have wild species suitable for domestication.[9] The result, says Diamond, was that Europeans arrived in western South Africa at a time when the Bantu-speaking Africans were not there, and so the Europeans had prior rights to the land. Diamond seems not to know that he is accepting here a large part of the historic myth used to justify apartheid.[10] Thus: an environmentalistic argument that supposedly explains a significant part of the history of South Africa down to modern times.

But the Khoi evidently did not adopt Xhosa agriculture for quite different reasons. Almost all of the area in South Africa that the Khoi occupied before the Europeans arrived is just too dry to support rain-fed agriculture; it is savanna and semidesert and has a subtropical climate, not a Mediterranean climate. The Khoi could have adopted irrigating and drained-field agriculture from their Bantu neighbors and farmed in a few seasonally wet riverside areas. They must have known about the Xhosa techniques of farming. But they chose to remain pastoralists. This had nothing to do with nondiffusion of Mediterranean crops and absence of domesticable plants. For pastoral peoples there is a conflict in timing between the grazing movement of herds and the needs of crops grown in a few widely scattered wetlands.[11] And changing from a pastoral economy to an agricultural one requires a rather drastic change in other dimensions of culture. The decision to retain a pastoral way of life was ecologically and culturally sound.

Actually, the zone of Mediterranean environment, with enough rainfall for cropping, is a quite tiny belt along the southernmost coast and adjoining rugged mountains, a region too small to bear the weight of argument that Diamond places on it. (His discussion of the Cape non-farming problem takes up no less than 25 pages of his book.) The Xhosa traded with the Khoi people in this region for products of animal husbandry, fish, seals, and the like; they had no incentive to displace the Khoi. There was extensive intercourse between Bantu speakers and Khoi throughout much of the savanna (summer-wet) regions of southwestern South Africa, as well as in Namibia and Botswana. Some Bantu speakers did indeed settle in part of the Khoi territory, and vice versa.

Australia also "cries out for explanation," according to Diamond. Why did Native Australians (so-called Aborigines) not adopt agriculture during the thousands of years that neighboring peoples to the north, in and around New Guinea, were farming? Again we are told that the explanation is a matter of environment and location. Diamond accepts the common view of cultural ecologists that the hunting–gathering–fishing economy employed by Native Australians was productive enough to give them a reasonable level of living so long as they kept their population in check (which they did). It is possible, also, that their way of life helped them to fend off efforts by non-Australians to settle northern Australia. Why, then, should they give up this mode of subsistence and adopt agriculture? Diamond simply assumes that they would have done so had it not been for environmental barriers.

It is true that most of Australia is desert and dry savanna, but the tropical north and northeast coast, the nontropical east and southeast coast, and a bit of southwestern Australia receive enough rainfall to sustain agriculture. But these regions, says Diamond, did not become agricultural because of their isolation from farming peoples outside of Australia. Diamond notes that Macassarese traded with Native Australians in the northwest, near modern Darwin, but he believes, oddly enough, that the Macassarese—who were famous sailors, by the way, and came from a region in Indonesia with productive agriculture—could not have sailed a mere 1,000 kilometers farther to the east, to the Cape York Peninsula, where tropical crops would have done quite well. But the Cape York Peninsula is itself very close to New Guinea, separated from it by the narrow Torres Strait, with several stepping-stone islands nearly connecting the two landmasses. Why did the Australians around Cape York not adopt the agriculture practiced by New Guineans? Again: isolation. Diamond argues, unconvincingly, that Australians would not have visited New Guinea itself during the thousands of years that (most of) the latter peoples practiced agriculture. This is hardly credible, and it seems to imply a belief that Native Australians were somehow less than rational. It is much easier to stick with the cultural–ecological argument that Australians chose not to adopt agriculture because they managed well without it.

The Americas pose a special problem for Diamond. He asks: why were New World peoples conquered by Old World peoples (Europeans) instead of the other way around? Why, in other words, did this hemisphere, much of which enjoyed the temperate climate that Diamond believes to be so critical for cultural evolution, not develop as rapidly as the Old World? There is a conventional scholarly answer to this question, and in fact it incorporates many geographical causes. The New World was not populated until recently in human history: perhaps 15,000–20,000 years ago. The people who arrived, in small numbers, from Siberia, were hunter-gatherers, not farmers.[12] At this time Old World cultures were beginning to experiment with agriculture. In the New World, the agricultural revolution began somewhat later, perhaps around 5000 B.C., and the level of sociopolitical complexity attained by 1492 was well behind that of the Old World. It is generally argued that the reason for this lag was the fact that hunting-gathering worked quite well for the Americans in this resource-rich and previously untouched environment until their population eventually reached the level where it would make sense for them to experiment with and then adopt agriculture in order to increase food supply. It is also argued by most scholars that there was no significant diffusion of culture traits from the Old World to the New during this entire period. The conquest of the New World resulted in part from its lower level of technology in 1492, but in much greater part from the susceptibility of Americans to Old World diseases to which, because of their long isolation, they had no immunity, and so they suffered devastating population losses. Diamond is not satisfied with this explanation, in spite of the fact that it incorporates arguments about isolation and diffusion.

Diamond's scientism leads him to pose historical questions in terms of universal principles of environmental causation. In essence: "Wherever we have A, we will have B; and if no A, then no B." Recall his argument about north–south versus east–west axes. He argues that all continental landmasses with an east–west axis will surpass all landmasses with a north–south axis. For this argument to be valid as a scientific generalization, it would have to explain all north–south cases; but there are only three: Africa, East and Southeast Asia with Australia, and the Americas. Moreover, he argues that tropical belts intervening between temperate regions will inhibit diffusion of agriculture (and everything else) between the temperate regions. Again there are three cases: Africa, the Americas, and the region extending from China south through Southeast Asia to Oceania. In each of these cases there are temperate regions at the northern and southern ends and a tropical belt in the middle. For Diamond, the most vexing of these cases is the New World. He wishes to explain the differences in levels of development in 1492 between Eurasia and the Western Hemisphere in terms of the same principles that he thinks apply to other regions, and thus show that the case for Eurasian superiority or priority applies to all other parts of the world, including the Americas.

Diamond therefore rejects the argument that the differences were caused by the lateness of New World settlement, leading to a late agricultural revolution. Instead, he argues, without evidence, that population growth in the New World was so rapid that arguments grounded in the recency of settlement and abundance of resources for hunting and gathering would be invalid; that the New World would have been on a social and technological par with the Old World in 1492—had it not been for the effect of environmental factors. There were, he says, four main noncultural reasons for Western Hemisphere backwardness in 1492. First, the Americas have a north–south axis. This must inhibit diffusion of cultural innovations between North and South America and later between the two regions of complex society (essentially Mexico and Peru). Second, the region lying between Mexico and Peru is tropical, hence a barrier for temperate-climate crops domesticated in each of the two regions. Third, North and South America are connected only by a narrow neck, the isthmus of Panama, and this inhibits diffusion. Fourth, diffusion northward from the Mesoamerican culture hearth into the temperate part of North America was rendered difficult, and was very slow, because, according to Diamond, the deserts of northern Mexico separate central Mexico from temperate North America.[13] One responds to the first two of these environmentalistic arguments with the same counterarguments that were offered in our previous discussion: the fallacies of north–south axes and tropical nastiness. The third argument is invalid because the width of the isthmus of Panama did not inhibit diffusion: there was sea travel, and there was movement of crops (notably corn) and other traits between the two continents.[14] And as to the fourth argument, it is simply bad geography. Diamond to the contrary notwithstanding, there is no desert separating northern Mexico from central and eastern North America—merely a savanna region intersected by waterways (central and east Texas), a region that could be, and was, crossed easily by diffusion processes.

The final part of Diamond's explanation for the agricultural superiority of Eurasia concerns domesticated animals. He is on somewhat firmer ground here when he stresses the priority of western midlatitude Eurasia, since many important species were domesticated in the region of grasslands, desert, open brushland, and forest extending from North Africa through the Near East into central Asia. Animal domestication played a lesser role than plant domestication in the origins of agriculture, so the Eurasian priority in this aspect of agriculture can be balanced off against other regions' priority in other aspects, such as Southeast Asia in rice and taro, tropical Africa in yams and sorghum, and so on. Moreover, although the Near East and adjoining North Africa and central Asia was the area of domestication of sheep, goats, horses, camels, and (probably) one species of cattle, India was the source of another species of cattle (*Bos indicus*), Southeast and south Asia that of water buffalo and (probably) pigs, South America of llama and alpaca, and so on. Cattle were herded in the region comprising the Sahara and the sub-Saharan Sudan as early as 7000 B.C., when rainfall in that region was much higher than it is today, and this may well be an area of domestication of one variety of cattle.[15] So it is more than an exaggeration for Diamond to say that "the successful [large animal] domesticates were almost exclusively Eurasian" (p. 157).

Diamond wants to show that Eurasia's importance in animal domestication was one of the primary reasons why temperate Eurasia (supposedly) gained superiority in subsequent cultural evolution. One argument is that large ungulates in tropical regions, for instance zebra, somehow were not suitable for domestication. But this is circular: Diamond can only show that those species that actually were domesticated were suitable for domestication. (He lamely argues that the failure of a small and brief nineteenth-century effort to domesticate zebra is evidence that this species could not be domesticated, when in fact domestication generally involved long time spans and a lot of work by many cultures.)

Diamond's crucial arguments about animal domestication concern the supposed implications and consequences of the process, and here he rehearses some familiar and erroneous theories. One claims that the horse revolutionized warfare, hence gave west-Eurasian (and especially Indo-European) horse-using warriors an advantage over all others, leading then to the development of complex societies first in this region; this is purely conjecture, and widely disputed. The use of horses and chariots in warfare may just as easily have been the consequence as the cause of early conquests. Diamond's contention that horses and cattle could not be used effectively in tropical Africa because of diseases such as trypanosomiasis is also invalid: disease-resistant varieties were widely employed in most (not all) parts of that region.[16] His claim that the domestication of cattle in western Eurasia explains the use of plows in this region is again invalid: plows were used very early in India (with cattle) and quite early in Southeast Asia (with water buffalo); plows are used elsewhere in the tropics (including Ethiopia); and the use of plows reflected the nature of farming systems: plowing generally is poor practice for most humid-tropical staple crops, Finally, Diamond's claim that the domestication of the horse and cattle in western Eurasia gave this region a great advan-tages in the transport of products, hence in the distribution of surplus production, is, again, invalid: draft animals came into use as a conse-quence of the development of surplus-producing agriculture, not as a cause of it. Animal domestication and animal husbandry were indeed im-portant for cultural evolution, but they gave no "ultimate" advantage to Eurasia.

CIVILIZATION

The "ultimate" environmental factors or forces, which caused agricul-tural societies to arise in some places and not others, continued to shape cultural evolution thereafter, according to Diamond. He discusses the evolution of writing, sociopolitical complexity, and technology, devoting most attention (unsurprisingly) to technology. Here is his summary of the argument about technological evolution after the Neolithic era:

> [Three] factors—time of onset of food production, barriers to diffusion, and human population size—led straightforwardly to the observed interconti-nental differences in the development of technology. Eurasia . . . is the world's largest landmass, encompassing the largest number of competing so-cieties. It was also the landmass with the two centers where food production began the earliest: the Fertile Crescent and China. Its east–west major axis permitted many inventions adopted in one part of Eurasia to spread rela-tively rapidly to societies at similar latitudes and climates elsewhere in Eur-asia. . . . It lacks the severe ecological barriers transecting the major axes of the Americas and Africa. Thus, geographic and ecological barriers to diffu-sion of technology were less severe in Eurasia than in other continents. Thanks to all these factors, Eurasia was the continent on which technology started its post-Pleistocene acceleration earliest and resulted in the greatest local accumulation of technologies. (pp. 261–262)

Diamond asserts, correctly, that people of all human groups are equally inventive. So he asks: what would lead to the piling up of the most inventions in certain areas, among certain groups, and hence to steady technological development in those areas? The broad answer is given in the passage quoted above. But we have seen that the "axes" are irrelevant, and the supposed "geographic . . . barriers to diffusion of technology" do not exist—or, rather, the barriers that chop up midlatitude Eurasia into separate agricultural regions are at least as significant as those between midlatitude Eurasia and tropical lands to the south.

What, then, is left of Diamond's explanation? Not very much. He supplies a brief and standard description of the way in which technology developed after Sumer and the way nonagricultural innovations spread westward to Europe and evolved in China. In this description he fails to mention the fact that diffusion eastward and southward from the Near East via the Indian Ocean, and southward from China via the South China Sea, was as important, and as easy, as diffusion westward. (Diffusion by way of India and the inner Asian land route is not discussed.) The next argument is a cracker-barrel environmentalistic theory about the things that supposedly lead to invention and innovation. In essence, the larger the population and the larger the number of so-called competing societies, the more inventions and innovations there will be. Therefore, since Eurasia is geographically the largest landmass, it will have the largest number of inventions and innovations. And they will diffuse through Eurasia's temperate belt more rapidly than they would in nasty tropical climates. Culture has nothing much to do with the process. Diamond uses roughly the same form of argument when he discusses the diffusion of writing and sociopolitical complexity from the Near East westward to Europe.

Nothing more needs to be said here about Diamond's account of human progress from Neolithic times to the present. My concern in this chapter is not with the problem of explaining cultural evolution after the Neolithic era. I wish merely to show that Diamond does not add anything significant to our understanding of this process by asserting the primacy of geography: of environment and location. Geography is important, but not *that* important.

EUROPE AND CHINA

Diamond's argument proceeds inexorably, deterministically, to the conclusion that Europe and China were fated to be the winners in the worldwide historical competition because of their environmental advantages. Europe was fated to be the ultimate winner, mainly because Europe's environment is superior to China's (recall the subtitle of the book: "The Fates of Human Societies"). History centers itself on temperate Eurasia, and the two regions of Eurasia that have the best environmental conditions for agriculture—for the origins of agriculture, and thereafter for food production—are Europe and China. Diamond sees Europe as the natural extension of the Fertile Crescent; the latter region lost out because of its lower ecological productivity, and history shifted westward to Europe. So we end up with just the two finalists: Europe and China.

China, says Diamond, dominated the eastern part of Eurasia as Europe did the western part. China's dominance began with the Neolithic era in north China. Diamond states as fact some extremely uncertain, in part quite dubious, archaeological hypotheses to argue that an agricultural revolution in central China led to the spread of farming peoples southward, displacing hunter-gatherer peoples in island Southeast Asia—thus, to show that there was here a north–south axis that had to favor temperate China at the expense of tropical Southeast Asia (and islands beyond). But it is by no means certain that farming is older in China than in Southeast Asia. Recall that the earliest known dates for agriculture in New Guinea are roughly as old as those for China, and recall how hard it is to get evidence on agricultural origins in humid-tropical regions like New Guinea. Moreover, rice was probably domesticated in India or Southeast Asia, not China, and may be as old as staple crops first domesticated in north China.[17] Diamond reinforces his argument with data from historical linguistics that provide what he thinks is solid evidence that all Austronesian (roughly, Malayo-Polynesian) languages derived originally from mainland China, via Taiwan. Indeed there is not much doubt that Austronesian languages originated somewhere in that region, but it could have been in any or all portions of the coastal regions stretching from south China down to Vietnam and Thailand—perhaps even from adjoining areas that were inundated by rising sea levels and are now shallow portions of the Sunda Shelf. (And Taiwan, be it noted, is also tropical, as is the south coast of China.) In sum, Diamond argues that China always had priority and centrality in all of eastern Eurasia, and history elsewhere in that region mainly reflects diffusions and migrations from a temperate China core.[18] This is mostly speculation, but Diamond's theory requires that it be true.

Finally we come to Europe. Most of the argument of *Guns, Germs, and Steel* is devoted to proving the primacy throughout history of midlatitude Eurasia, and within this region of Europe (supposed heir to the Fertile Crescent) and China. If the argument stopped there, we would have a sort of Eurasia-centrism, not Eurocentrism.[19] But Diamond's purpose is to explain "the broadest patterns of history," and so he must answer this final question: why did Europe, not Eurasia as a whole, or Europe and China in tandem, rise to become the dominant force in the world? Diamond's answer is, predictably, the natural environment. The "ultimate" causes of Europe's rise, relative to China, are a set of qualities that Europe's environment possesses and China's environment lacks, or that China possesses, but in lesser degree. The "ultimate" environmental causes then produce the "proximate" causes—which are cultural:

> [The] proximate factors behind Europe's rise [are] its development of a merchant class, capitalism, and patent protection for inventions, its failure to develop absolute despots and crushing taxation, and its Graeco-Judeo-Christian tradition of empirical inquiry. (p. 410)

This is, of course, utterly conventional Eurocentric history. We have discussed this model at length in earlier chapters of this book and in Volume 1, and I do not need to repeat here the reasons that I consider it to be invalid. I do need to point out, however, that there is now a huge literature that systematically questions each of these economic, political, and intellectual explanations for the rise of Europe, much of this literature consisting of Eurocentric arguments of one sort attacking Eurocentric arguments of some other sort—yet Diamond ignores all this scholarship and simply announces that these (and a few other cultural things) are the true "proximate" causes of the rise of Europe. Evidently Diamond views the matter as settled. The problem, for him, is to find the underlying environmental causes.

Topography is the key—or, more precisely, topographic relief and the shape of the coastline.

> Europe has a highly indented coastline, with five large peninsulas that approach islands in their isolation. . . . China's coastline is much smoother. . . . Europe is carved up . . . by high mountains (the Alps, Pyrenees, Carpathians, and Norwegian border mountains), while China's mountains east of the Tibetan Plateau are much less formidable barriers. (p. 414)

These observations about physical geography—they are rather inaccurate, as we will see—lead into one of the truly classical arguments of Eurocentric world history: the theory of Oriental despotism.[20] This is the belief that the so-called "Oriental" civilizations—essentially China, India, and the Islamic Middle East—have always been despotic; that Europeans alone understand and enjoy true freedom; that Europe alone,

therefore, has had the historical basis for intellectual innovation and social progress. Back in the seventeenth and eighteenth centuries, the
seemingly obvious fact that Europeans alone knew freedom was attributed (by Europeans) to the fact that they alone believed in the True God.
After the mid-nineteenth century, European historians invoked secular
causes, namely the three foundation arguments that were mentioned at
the beginning of this chapter: race, essential culture, and environment.
Most of the Eurocentric historians discussed in this book invoke both
culture and environment as complementary causes, For instance, Michael Mann (Chapter 6) thinks that the ancient Greeks invented rational thought and the ancient Germanic tribes invented the love of freedom; thereafter, true European freedom, democracy, individualism, and
so on, were brought to fruition thanks mainly to Europe's uniquely favored environment. Jared Diamond looks to the environment as the true
cause and dredges up a pair of old environmentalistic theories, adding
nothing new to them, about how physical geography is the main reason
why Europe, not China, acquired the cultural attributes that gave it ultimate hegemony: "a merchant class, capitalism . . . patent protection for
inventions . . . failure to develop absolute despots and crushing taxation,"
and so on.

Here is how it works, according to Diamond. China is not broken up
topographically into isolated regions because it does not have high
mountains like the Alps and does not have a coastline sufficiently articu-
lated to isolate nearby coastal regions from one another. This explains
the fact that China became unified culturally and politically 2,000 years
ago. Europe, on the other hand, could not be unified culturally and politically because of its indented coastline (its "capes and bays," in the traditional theory) and because of its sharply differentiated topographic relief
(its "many separate geographical cores," in the traditional theory). Europe therefore developed into a mosaic of separate cultures and states.
China's geographically determined unity led it to became a single state,
an empire; and an empire must, by nature, be despotic. Why? Because a
person cannot leave one state and emigrate to another to avoid oppression, since there is only the one state, the Chinese empire. Hence, there
is continued oppression of the populace and centralized manipulation of
the economy. So: no freedom, little development of individualism, little
incentive to invent and innovate (taxation, political control, and so on),
no development of free markets, and no development of a polity resem-
bling the modern democratic nation-state. These "harmful effects of
unity" (p. 413) led China to begin to fall behind Europe some five hun-

dred years ago. Diamond concedes that China had indeed been innovative in earlier times; it had even begun to move toward an industrial revolution during the early Middle Ages; but China's oppressive imperial
despotism led it to stagnate after the fourteenth or fifteenth century. Europe, by comparison, continued to forge ahead. Therefore Europe triumphed.

The geography is wrong and so is the history. Southern Europe has
the requisite "capes and bays" (or peninsulas) and separate "geographic
cores." But the historical processes that Diamond is discussing here per-
tain to the last 500 years of history, and most of the major developments
during this period, those that are relevant to his argument, occurred
mainly in northern and western Europe, which is rather flat: the North
European Plain from France to Russia; the extension of that plain across
France almost to the Spanish border; southern England (which is barely
separated from the European mainland). Even central Europe is not really isolated from northern and western Europe. There are few significant
coastline indentations between Bordeaux and Bremen. If we look at the
distribution of population throughout this region, there is no isolation
and not very much development of cores. The crystallization of northern
Europe's tiny feudal polities into modern states occurred for reasons that
had little to do with topographic differentiation; the boundaries of most
of these states do not reflect topographic barriers, and most of their cultural cores are not ecological cores. The idea that the pattern of multiple
states somehow favored democracy is a misconception: each of these
states was as despotic as—probably much more despotic than—China,
and emigration from one polity to another was not substantial enough to
have had any effect on the development of democracy. Further, what Diamond calls (euphemistically) Europe's "competing" states often were
warring states; probably China was more peaceful during most centuries
than Europe was (perhaps even during the Ming–Qing disruptions), and
an environment of peace surely is more conducive to economic development than one of war. And finally, Diamond's view of Chinese society is
based on outdated European beliefs. China did not stagnate in the late
Middle Ages: Chinese development continued without interruption, and
Europe did not outdo China in technology, in the development of market institutions, and indeed in the ordinary person's standard of living
until the eighteenth century.[21] In short, the idea that China's topography
led to China's achievement of a unified society and polity, and that this
unity somehow led to despotism and stagnation, is simply not supported
by the facts.

Diffusion is also supposed by Diamond to have played a large role in the triumph of Europe over China. Throughout *Guns, Germs, and Steel*, Diamond argues that geographical barriers to diffusion are one of the main reasons why some societies failed to progress. But China, he argues, had fewer barriers to diffusion than Europe had. Should China, therefore, have progressed more rapidly than barrier-ridden Europe? How does he get around this contradiction? First, he introduces a tortuous theory to the effect that, not only is too little diffusion a hindrance to development, but so, too, is too much diffusion. Like the second of the Three Bears, Europe had just the right balance between too little differentiation and too much, and this, mysteriously, led to more intense diffusion of innovations in Europe than in China. Second, Diamond claims—another traditional argument—that Europe's lack of political unity somehow favored the diffusion of innovations, whereas surely it did the opposite. Political boundaries are barriers to human movement; also, they frequently correlate with linguistic boundaries and thus can be barriers to communication. The third argument is largely an implicit one, though clearly evident nonetheless. Diamond claims that social and technological development moved steadily westward from the Fertile Crescent to Europe. He states (incorrectly) that writing, invented in the Fertile Crescent, was merely a tool of the ancient despotic bureaucracies until the alphabet diffused westward to Greece, where, he says (again incorrectly), the Greeks added all the vowels (not just some of them) and thereby transformed it into an instrument of creative writing, of innovation, abstract thought, poetry, and the rest. In essence, this is an argument that intellectual progress diffused westward and became consequential when writing reached Europe. This must be the basis for his argument that "the Graeco-Judeo-Christian tradition of empirical inquiry" is one of the reasons why Europe triumphed. Yet, throughout *Guns, Germs, and Steel* Diamond insists (rightly) that all peoples are equally creative, equally rational. This is a contradiction; but in fact it is a nonissue, since "empirical inquiry" was not invented by Europeans and was as highly developed in China, and other civilizations, as in Europe.

I described Diamond's argument as "scientistic," not because he tries to use scientific data and scientific reasoning to solve the problems of human history. That is laudable. His argument is scientistic because he claims to produce reliable, scientific answers to these problems when in fact he does not have such answers, and because he discards wholesale the findings of social science while inserting old and discredited theories of environmental determinism. That is bad science.

NOTES

1. Diamond, *Guns, Germs, and Steel: The Fates of Human Societies* (1997).

2. Parenthetical page numbers in this chapter refer to *Guns, Germs, and Steel.*

3. Davidson, *The Prehistory of New Zealand* (1984); King, *Moriori: A People Discovered* (1989); Bulmer, "Gardens in the South: Diversity and Change in Prehistoric Maaori Agriculture" (1989).

4. See Lewis and Wigen, *The Myth of Continents: A Critique of Metageography* (1997), for a fine discussion of the arbitrariness of the notion of continents.

5. The idea that the axes of continents have had a profound effect on world history is an old one; see, for instance, in Karl Ritter's 1865 book *Comparative Geography* the section that is titled "The Position of the Continents and its Influence on the Course of History" (pp. 46 *et seq.*). Ritter believed that God shaped the continents to benefit humanity.

6. Glover and Higham, "New Evidence for Rice Cultivation in South, Southeast, and East Asia" (1996).

7. This argument used to be a popular explanation, in "modernization theory," for the fact that humid tropical regions are underdeveloped: the problem is bad eating habits. As one paleobotanist, Jack Harlan, points out, "One can more or less live on potatoes if one eats enough of them"; in Harlan, *The Living Fields: Our Agricultural Heritage* (1995), p. 190. Also see Blaut, "The Ecology of Tropical Farming Systems" (1963).

8. The Inner Asian Silk Road, through the deserts north of Tibet, was developed too late to bear on these issues. Historians sometimes assume that the earlier movements of pastoral peoples in this region led to east–west diffusion of staple crops; but these people were not farmers, and oases were few and far between.

9. Blumler, "Ecology, Evolutionary Theory, and Agricultural Origins" (1996), p. 36, disputes the claim that wild species that were potentially staple domesticates were not available in the Cape region.

10. Diamond concedes that the Khoi people would have had the best claim of all, but they were exterminated or driven out by the Europeans and therefore don't count. He might have added that the Europeans defined the Khoi pastoralists as nomads, hence people who had no legitimate claim to any territory—a classic example of the colonialist "myth of emptiness" (see Volume 1).

11. Khoi people in fact lived among the Xhosa in some areas. Perhaps we will find that Khoi pastoralists practiced crop farming on the side or in areas close to Xhosa settlements.

12. The people called "hunter-gatherers" were also, in many areas, fishers and shellfishers.

13. In addition, Diamond agrees with everyone else that lack of immunity to Old World diseases brought by the Europeans was a factor in the conquest.

14. Fiedel, *Prehistory of the Americas* (1987).

15. See Blench, "Ethnographic and Linguistic Evidence for the Prehistory of African Ruminant Livestock, Horses, and Ponies" (1993).

16. Giblin, "Trypanosomiasis Control in African History: An Evaded Issue?" (1990); Turshen, "Population Growth and the Deterioration of Health: Mainland Tanzania, 1920–1960" (1987); Shaw, *The Archaeology of Africa* (1993).

17. MacNeish, *The Origins of Agriculture and Settled Life* (1991); Glover and Higham, "New Evidence."

18. Diamond ascribes the fact that Korea has a language very different from Chinese to its "geographical isolation" from China, but there is no such isolation—another contradiction in Diamond's treatment of diffusion and nondiffusion.

19. Eurasia-centrism has achieved a certain popularity in the past few decades. This reflects the new scholarship that demonstrates the historical achievements of China and also the present-day economic success of Japan. Eurasia-centrism has not supplanted plain Eurocentrism, of course.

20. We discuss various forms of the theory of Oriental despotism in Chapters 2, 5, and 9. Diamond does not call his theory "Oriental despotism," but that is what it is.

21. See in this regard Frank, ReORIENT (1998); Wong, China Transformed: Historical Change and the Limits of European Experience (1997); various essays in Twitchett and Mote, The Cambridge History of China, Vol. 8: The Ming Dynasty (1998); The Colonizer's Model of the World, Volume 1; and Chapters 4 and 5 in this volume.

CHAPTER 9

David Landes:
The Empire Strikes Back

David Landes's 1998 book on world history, *The Wealth and Poverty of Nations: Why Some Are So Rich and Some So Poor*, was enthusiastically reviewed in *The Wall Street Journal*, *The New York Times*, and the *Washington Post* before it even reached the bookstores.[1] When this kind of attention is given to a history book, we tend to suspect that it says something that the opinion makers of our society very much want us to believe. This book strikes back at the critics of Eurocentric history. They are guilty, says Landes, of "Europhobia" (p. 514). They are, basically, ideologues, for whom "the very idea of a . . . Eurocentric global history is . . . arrogant and oppressive" (p. 513). They aim "to shape the truth to higher ends" (p. 348). Scholarly history, says Landes, *should* be Eurocentric: "Some say Eurocentrism is bad. . . . As for me, I prefer truth to goodthink" (p. xxi).

Landes's "truth," however, looks very ideological. For instance: "[The] very notion of economic development was a Western invention" (p. 32). "Over the thousand and more years of . . . progress . . . the driving force has been Western civilization and its dissemination" (p. 513). "Sub-Saharan Africa threatens all who live or go there" (p. 8). African farmers prefer large families as "proof of virility" (p.501). "Chinese lacked . . . curiosity" (p. 96). "*Chinese savants had no way of knowing when they were right*" (p. 344, emphasis in the original); "unlike China, Europe was a learner" (p. 348). In Latin America, "the skills, curiosity, initiatives, and civic interests of North America were wanting," and "independence slipped in [as] a surprise to unformed, inchoate entities that had no aim but to change masters" (p. 313). Japanese have exhibited a "characteristic ferocity" (p. 355). Indians (before British rule) were "a docile people" (p. 396). "Even if Israel did not exist, [the Arabs] would be at one another's throats" (p. 409). And so on. All of these quotes are of course taken out of context, but they convey the flavor and tone of *The Wealth and Poverty of Nations*.

Ordinary Eurocentrism is not exactly hot news, however. What must have excited *The Wall Street Journal* in particular was Landes's argument that, basically, the history of the world is the history of the West and the history of the West is a march toward an end condition, "the ideal growth-and-development society" (p. 216), which looks very much like pure laissez-faire capitalism. Also exciting was Landes's argument that imperialism is natural: it is "the expression of a deep human drive" (p. 63) and a good thing, while modern globalization, structural adjustments, and the like are also natural and good. These arguments are very useful to those who, so to speak, want to freeze history right where it is here and now.

In this chapter I will analyze Landes's theory of history in considerable detail and show that—borrowing his words—it is not truth but Eurocentric goodthink. The analysis will roughly follow the sequence of arguments in *The Wealth and Poverty of Nations*. The first argument (Chapters 1 and 2) claims that Europe, especially northwestern Europe, has a natural environment that is superior to all other places in the world. The second argument (Chapters 3 and 4) holds that European culture, particularly the European mind, has been superior to all other cultures and mentalities since Old Testament times. The third argument (Chapters 5 through 12) deals with the history of European expansion, imperialism, and colonialism, explaining why, in Landes's view, the process was both natural and good. The fourth argument (Chapters 13 through 19) is an effort to explain the industrial revolution in Eurocentric terms. The fifth and final argument (Chapters 20 through 28) is a look at the major societies of the world today, showing, supposedly, why Europeans and the European way are superior to all other peoples and their various modes of production and life, and why, for their own sake, these other societies should knuckle under to the Europeans.

"WARM WINDS AND GENTLE RAIN"

We noted in Chapter 8 that two basic foundation theories underlie most modern arguments for the historical superiority of Europe: a better natural environment; more progressive culture.[2] Landes uses both, starting with the environment.

What we are offered is classical environmental determinism. It needs to be said at the outset that geographers have long since discarded this theory.[3] Unsurprisingly, Landes resurrects the early-twentieth-century views of Ellsworth Huntington concerning the supposedly determining influence that climate has on human activity.[4] Huntington was the most famous exponent of what we are calling in this book "Euro-environmentalism," that, is, environmental determinism deployed in the service of Eurocentrism. Following Huntington, Landes argues that tropical climates are inimical to human activity and cultural progress. Why so? Landes gives a series of supposed reasons, each of which I will show to be false.

Landes begins by pointing to the map and asking us to notice that rich countries tend to be located in "temperate" (midlatitude) regions and poor countries in the tropics. Landes asserts that this is causation, not merely correlation: tropical climates are bad for human progress. Actually, any historical theory that explains the fact that Europe began to "rise" after 1500, and thereafter became richer than all other societies, will serve quite nicely to show why countries in temperate regions are on the whole wealthier than countries in the tropics. Europe's development did not just diffuse outward in all directions. Europeans settled in regions that allowed them to practice familiar farming systems, and from this agricultural base developed outliers of European (British) society in these temperate regions. Anglo-America and also Australia–New Zealand have been integral parts of a single economy that was centered on Britain until the late nineteenth century. Temporary disruptions, like the American Revolution, have not really altered this fact. Stated in another way: the relationship between Britain and Anglo-America has not been that of imperial core and exploited periphery; it has been that of two essentially equal parts of a single system. By contrast, all of the rest of the world has been, from the British (and Dutch, and French, etc.) point of view, hunting grounds for profit. Sugar and cotton were by far the most profitable agricultural commodities down to the early nineteenth century; both are essentially tropical and subtropical crops; and so there developed a plantation economy controlled by Europeans, exploiting the (emptied) land of Latin America, and seizing labor for the plantations from the only nearby center of dense population, West Africa. In Asia, nontropical China and Japan were too remote to be brought into the Europe-centered economy until well into the nineteenth century; China then began to become underdeveloped, while Japan, because it was even more remote from European military power, successfully resisted European imperialism. All of the foregoing summarizes what was discussed at greater length in Volume 1. So my point here is simply this: the underdevelopment of tropical regions is a consequence of history, not climate.

One of the premises of classical environmental determinism was the idea that heat, somehow, impedes human mental and physical activity. Landes repeats this argument, apparently unaware that it has been discredited. Again following Huntington, he asserts that a kind of medium, or "temperate," temperature regime is better for humans than one that is too hot or too cold. But, he says, too much cold can be resisted with clothing and shelter, but not so too much heat. In fact, we know that human bodies accustomed to relatively hotter surroundings can function just as well as those accustomed to colder surroundings.[5] Landes sees climate as part of the explanation for slavery: Europeans could not work under the hot sun, so it was somehow natural to force Africans to work on the plantations.[6] (This was a favorite argument of proslavery publicists in the old days. For Landes, it fits into his larger theory absolving Europeans of essentially all blame for the negative effects of imperialism.) This is a fallacy. There are many tropical regions, Queensland being one, in which Europeans have traditionally undertaken field labor; and we may note that, in the semitropical Southern United States, white farmers after the Civil War worked the same fields as black slaves did previously, and some still do so. Be it noted, also, that farmers in the humid tropics, where there is no winter, can work their fields year-round. Few farmers would agree with Landes that "winter . . . is the great friend of humanity" (p. 8).[7]

Landes then asserts that people in tropical climates are plagued with diseases. In fact, people in poor countries in general are plagued with diseases, and the reason is poverty, not climate. It is true, as he says, that cold weather suppresses insect vectors for some diseases, but this is only one of many relevant environmental variables.[8] Mammalian hosts are now known to be the main pools of infection for many human diseases, and domestic animals (along with rats) are as important in this regard in midlatitudes as in the tropics. Many of the so-called tropical diseases used to plague midlatitude regions: malaria, for instance, used to be a curse in New York City. Landes focuses on tsetse flies and trypanosomiasis, and recites the old colonial-era falsehoods about this disease in Africa. "Tsetse makes large areas of tropical Africa uninhabitable by cattle and hostile to humans. . . . Animal husbandry and transport were impossible." (p. 9) This just isn't so. In fact, it seems now fairly likely that the tsetse fly problem was largely controlled in Africa—perhaps as much as anthrax was in Eurasia—until the slave trade depopulated large areas, leading to a great expansion of bush, which vastly increased the population of wild-animal hosts.[9]

Landes rounds out the indictment of the tropics with several more false assertions. "Water is another problem in the humid tropics. . . . The timing [of rainfall] is often irregular . . . the rate of fall torrential" (p. 18). In fact, rainfall variability is a problem in all semi-arid regions, tropical and nontropical (though *not* the wet tropics), likewise, torrential downpours. (By the way, the worst winter storms in northern Europe are as fearsome as hurricanes are in the tropics.) Landes claims, again falsely, that food supply problems are results of these difficulties. This, he says (falsely), stems from the fact that tropical agriculture is shifting (so-called "slash-and-burn") agriculture that is hopelessly unproductive. In fact, most farmers in the tropics practice sedentary, not shifting, agriculture. Tropical soils are not—Landes to the contrary notwithstanding—infertile.[10] Food problems usually are problems of poverty, not environment.[11] But Landes's conclusion is simple, old-fashioned climatic determinism:

Life in poor climes . . . is precarious, depressed, brutish. (p. 14)

There are "far more favorable conditions in temperate zones; and within these, in Europe above all; and within Europe, in western Europe first and foremost" (p. 17). Again we are given a litany of old and discredited environmentalistic arguments. Winters in western Europe are mild; "Europeans were able to grow crops year round"; but "mild" is just a value judgment, and winter cropping (aside from perennials) was possible only in small areas of southern (not western) Europe (p. 17). Western Europe had "warm winds and gentle rain, water in all seasons, and low rates of evaporation" (p. 18). In fact, the climate in much of this region, particularly in the northwest, is so wet that solar energy is limited, grain crops sometimes cannot do well, and soils generally do not dry out until late spring if at all.[12] Says Landes: in eastern Europe winters are more severe, while in southern Europe rain is sparser; and—another classical fallacy—all of this led to greater poverty and less industrialization in eastern and southern Europe than western Europe. Europe's livestock are sturdier and healthier than those of other regions, thanks to the climate, according to Landes. (Not so.[13]) Other Euro-environmentalistic false beliefs are added; these, for the most part, are borrowed from Eric L. Jones's *The European Miracle* and are refuted in Chapter 5 of this volume.

Landes wants to proclaim the advantages of temperate regions in general, so he must deal with China.[14] He compares China unfavorably with Europe in various contexts throughout *The Wealth and Poverty of*

Nations, mostly disparaging the supposed irrationality of the Chinese in such things as economic, political, and sexual ("reproductive") behavior, matters we discuss later in this chapter. But Landes invokes one classical environmentalistic theory as part of the explanation for China's inferiority to Europe throughout history. This is the theory of "Oriental despotism," according to which civilizations centered on river valleys and based in irrigated agriculture supposedly are inherently despotic and unprogressive. We discussed this theory in Chapters 2, 5, and 6, and showed, I hope, that the theory is nonsense.

A UNIQUELY PROGRESSIVE CULTURE

The most striking thing about *The Wealth and Poverty of Nations* is the sheer number of different arguments that the author makes to explain the historical and present-day superiority of Europe. The arguments, all very casually stated, extend across the entire spectrum, from climate to culture, politics, economy, society, and mentality, and Landes seems to be telling the reader that any one of these arguments alone would be sufficient to explain European superiority. The book reads like a legal brief. However, the arguments are broadly organized into the two categories, environment and culture. We turn now to culture.

There is a rough order to these arguments about culture. First comes a chapter surveying essentially all aspects of society throughout all of history. Then comes a chapter focusing on the European mind. In both chapters we are mainly told about the superior qualities of Europeans, with some comparative discussion of the inferiority of non-Europeans. Later in the book come chapters devoted more specifically to denigrating non-Europeans: the unprogressive features of Islam, India, China, Africa, and Latin America. I will discuss these arguments broadly in the order they are presented.

There has always been a fundamental difference between East and West, says Landes. He then lists, but does not really discuss, a truly remarkable number of old Eurocentric beliefs about how and why the West has been more progressive, more enlightened, than the East throughout history. It should be said that most of these old beliefs, or theories, are still being advanced by some Eurocentric historians, including several of those whose views we discuss in this book. But it is important to realize that literally all of these theories have been challenged by scholars—yet Landes blithely announces these theories as though they were accepted facts.[15]

Eastern civilizations in ancient and medieval times were

> oriental despotisms (servitude for all). . . . What did ordinary people exist for, except to enhance the pleasure of their rulers? Certainly not to indulge a will of their own. . . . The very notion of economic development was a Western invention. Aristocratic (despotic) empires were characteristically squeeze operations: when the elite wanted more, they did not think in terms of gains in productivity. . . . They simply pressed (and oppressed) harder. (pp. 31–32)[16]

Setting aside for the moment the question whether this splenetic assertion applies any more to the ancient East than it does to the ancient West, we must offer some refutations. It is now fairly widely accepted that Eastern civilizations developed economically at rates, overall, that were no less rapid than the West. So did their technology. And living standards were no lower in Asia than in Europe until modern times; probably until the mid- or late eighteenth century. The so-called despotisms usually left peasants and merchants pretty much alone (if they paid their tribute). The ruling classes made up only a relatively small fraction of the population, and their exploitative activities, at most times and places, could not have put such heavy burdens on ordinary people as Landes depicts.[17]

Western civilization, according to Landes, took an entirely different course. We are told first of all that private property rights are inherent in the Judeo-Christian tradition,[18] as is also the idea of personal freedom (coauthored by the ancient Greeks).[19] Rome was a temporary setback: a despotic empire.[20] Christian teachings and, more important still, the rough individualism of the Germanic tribes further advanced the uniquely European ideas of freedom and private property. There is a strongly teleological flavor in all of this: democracy and capitalism were somehow foreordained in the ancient and medieval European past—and nowhere else.

This is classical Eurocentric mythology. All of these arguments were put forward by Eurocentric historians discussed in prior chapters, and were there refuted, so here a few brief comments will suffice. First, recognition of some limited property rights was general throughout ancient civilizations, although everywhere these rights were circumscribed in various ways and were overlain by some notion of the ruler's ultimately supervening right. The West was not unique. Second, there is no good evidence that personal freedom was recognized more fully in the West than in the East. Greek so-called democracy was merely equality among elite males.[21] And third, the culture of the Teutonic tribes of Europe had no more inherent respect for personal freedom and property than would be found in countless other tribal cultures: the argument is a traditional European cultural conceit.[22]

Landes then contrasts the evil empires of the East with what he sees as evolving democratic states of the ancient and medieval West. Again we are given one of the standard tales of Eurocentric history. (Pretty much the same tale was recounted by Eric Jones, Michael Mann, and John Hall, as we discussed in previous chapters.) At the root of this argument is the belief that the idea and enjoyment of freedom have been monopolies of the West for thousands of years. Then we are told that the fragmentation of European polities during the feudal era somehow meant that these political units were inherently more democratic than the huge empires of Asia. But, in reality, these entities were not really states: they reflected the fractured and layered sovereignty of that period, with feudal estates lying within larger entities, up to the level of the baron or king.

Landes's error, like that of the other Eurocentric historians who use this argument, is that of anthropomorphizing the feudal polity: somehow the jigsaw puzzle on the map is transformed into a population of human individuals, free and equal. Beneath this anthropomorphism is the hidden notion that the lords of these feudal polities themselves *were* the polities, and their gracious behavior toward one another—all within the feudal ruling class, of course—somehow implied democracy in the political entities that they ruled.[23] But there is no evidence that the political systems within feudal polities were more democratic than the systems to be found in the empires of Asia. Landes describes the latter as vicious despotisms, whereas the reality is quite different. Recent scholarship on China has shown that the Chinese empire allowed a considerable degree of freedom to its subjects; this, indeed, was one of the keys to its longevity.[24] There is no reason to believe that medieval European polities offered more freedom than did the empires of the East.

Next we are given the classic Weberian proposition that urbanization in medieval Europe was unique among cities of the world and was a crucial factor in Europe's political progress toward modern democracy. Weber, as we saw in Chapter 2, thought that European cities were more autonomous than Asian cities in the Middle Ages, but this view is contradicted by modern studies of Asian cities, which were pretty much on a par with European cities in economic and political terms.[25] Landes outdoes Weber: European cities conferred "civil power . . . social status and political rights on its residents. . . . [The] cities were gateways to freedom, holes in the tissue of bondage that covered the countryside" (p. 36). This is mythic: only a tiny fraction of serfs in Europe were able to flee to the cities, and the cities were generally hostile to in-migration of strangers. It is indeed true that city dwellers enjoyed more rights than did tied peasants, especially serfs, in the feudal countryside, and that the urban economy advanced technologically more rapidly than did the rural economy, but this was as true in Asia (and Africa) as it was in Europe. Landes then argues that European cities were natural allies of the monarchs in their progressive struggle against feudal landlords—a process that we know occurred as feudalism was ending and giving way to early-modern absolutist kingdoms—but comparable processes were under way elsewhere, and in any case it is hard to follow an argument that portrays European feudal lords as agents of progress and at the same time condemns them as opponents of progress ("tissue of bondage" and so on)—you can't have it both ways.

THE INVENTORS OF INVENTION[26]

Landes has been arguing for many years that the technological inventiveness that we know to be characteristic of modern industrial society was actually proceeding at full speed in Europe (but nowhere else) during the Middle Ages. He might be described as a member of the historical school of Eurocentric technological determinists, were it not for the fact that Landes, eclectically, invokes a number of different determinisms (including, as we saw, environmental determinism), more or less as each is needed to reinforce one or another of his Eurocentric arguments. Landes has, himself, done scholarly work on the development of certain technologies in medieval Europe; the trouble is, here, as elsewhere, he just fails to see that the advances that took place in medieval Europe were paralleled by advances taking place in China and other societies, sometimes in the same technical field, sometimes in related ones. China, India, and the Middle East were either on a par with Europe in overall technology and technological progress throughout the Middle Ages or (as some scholars argue) were actually in advance of Europe until at least the sixteenth century.[27] Landes believes otherwise.

Europeans have been the most inventive people on earth for millennia, according to Landes. He depicts an open, progressive, innovative society with roots that go back thousands of years, to biblical beginnings, to ancient Greece, and to prehistoric Germanic culture. But the point is made mostly by invidious comparisons with the other early civilizations, which are almost invariably characterized as unprogressive. It seems that

Asian civilizations were blocked from inventiveness and innovativeness by (1) Oriental despotism and (2) irrational reproductive behavior. Oriental despotism goes back all the way to the ancient riverine civilizations, supposedly characterizing China, India, and the Middle East throughout their history.

> As for China . . . the mandarinate and imperial court . . . stifled dissent and innovation, even technological innovation. This was a culturally and intellectually homeostatic society. . . . As soon as . . . change threatened the status quo, the state would step in and restore order.[28]

This is nonsense. China simply was not a "culturally and intellectually homeostatic society."

Landes adopts the traditional Malthusian view of overpopulation: in essence, non-Europeans are not rational enough to control their sexual urges and therefore have more children than they should, so that population grows out of control. Thus Asia supposedly had "early and almost universal marriage, without regard to material resources. . . . In contrast, Christian and especially western Europe accepted celibacy, late marriage (not until one could afford it) and more widely spaced births."[29] The "long-standing reproductive strategy" of the Chinese was "early, universal marriage and lots of children. That takes food, and food in turn takes people. Treadmill. This strategy went back thousands of years" (p. 23). A feeble effort is made to explain all this in a nonprejudiced way: "the demand for labor in the rainy season and the big yields of wet cultivation promoted high densities of population" (p. 21). But density of population tended to be roughly proportional to agricultural productivity in historical peasant societies: there is no logic to the argument that people who practice irrigated agriculture should have less reproductive restraint than people who practice any other form of agriculture if both forms yield the same amount of food per person. In any case, agricultural (and other) technology developed very nicely in China, and elsewhere in Asia. And recent scholarship tends to suggest that Europe did not differ from non-Europe in family size, birth rate, and the like, down through the Middle Ages.[30] So Landes's model falls prey to empirical evidence.

The period A.D. 1000–1500 in Europe saw "an economic revolution . . . such as the world had not seen since the . . . Neolithic."[31] This, says Landes, was the period when the unique inventivesss of Europeans truly came to flower. The initiating conditions were the uniting of Christianity and Germanic culture at the end of the Dark Ages.[32] Just how this amalgamation took place is not spelled out, but here, as elsewhere,

Landes offers a sort of contrapuntal theory, an invidious comparison with China and (especially) the Islamic Middle East. These civilization earlier had a potential for growth. China (says Landes) lost momentum around the end of the Song period, apparently because of the old afflictions of Oriental despotism and overpopulation, and thereafter began to regress, scientifically, technologically, and in many other ways. In the Middle East, progress was snuffed out by the Islamic religion. Landes paints Islam and the Islamic peoples in very dark colors. The Muslim world was progressive until about A.D. 1100 "Then something went wrong. Islamic science . . . bent under theological pressures for spiritual conformity. . . . For militant Islam, the truth had already been revealed."[33] But this view of medieval Chinese and Islamic society has been thoroughly discredited by scholars: we now know that science and technology flourished in these civilizations throughout the medieval period.

A characteristic error runs through Landes's discussions of medieval European technology. Like many other Eurocentric historians, Landes describes modern or early-modern traits and claims to find them appearing several hundred years before they actually did appear, using as evidence some primitive forerunner of the trait. Conversely, for non-Europe, he describes very early, primitive traits and claims that they are permanent characteristics of these societies. All of this makes medieval Europe seem more modern than it was and medieval non-Europe seem more backward.[34] The error is made both for attitudes—inventiveness and innovativeness especially—and for material traits; for the latter, some crude forerunner of a technique in medieval Europe is magically transformed into the later, fully developed form; and in societies like China, the early appearance of a fully formed technique is either denied (ignoring the known evidence) or is assigned some nonrational function. Example: Landes concedes, as he must, that gunpowder appeared earlier in China than in Europe, and cannons at least as early (in the late thirteenth century), but for China early cannons are described as inefficient, irrational. "They were apparently valued as much for their noise as for their killing power. The pragmatic mind finds this . . . vision of technology disconcerting." Then Landes leaps over three centuries of world history, to sixteenth-century Europe, and announces that Europe has "the world's best cannon" (p. 53). This example also exhibits another characteristic error: when a trait was invented outside of Europe, the emphasis is laid, not on the invention, but on the way "pragmatic" Europeans improved it.

Farmers in medieval Europe were both more modern and more progressive than their counterparts elsewhere, according to Landes. (Indeed, he reserves the label "peasant" for non-European farmers.) Land tenure was essentially private property; elsewhere it was not. Attention is focused on medieval England, and we are told of the "individualism" of English "yeoman farmers," as contrasted with "the dumb submission of the Asian *ryot.*" "Englishmen were free and fortunate."[35] And so on. Thus feudalism in all its barbarity is simply removed from the medieval rural landscape. (And whatever happened to that "tissue of bondage" that, he says, "covered the countryside"?) Moreover, the medieval western European rural landscape is being transformed by a technological revolution. Landes lists a number of supposedly revolutionary traits in agriculture, broadly following Lynn White, Jr. (see Chapter 3). He fails to inform his readers that this notion of a medieval technological revolution in agriculture is highly controversial among mainstream historians and that there is no reason to believe that change was occurring more rapidly in Europe than elsewhere.[36] As we saw in our discussion of White, the heavy plow was not exclusively a European invention and was introduced into northern Europe as an adaptation to the heavy soils of that region; the use of draft animals was neither new nor peculiarly European; the three-field rotation was hardly a revolutionary European innovation (it was known elsewhere). The same judgment holds for windmills, water wheels, irrigation and drainage systems, and essentially all of the other technological advances in medieval European agriculture: they were paralleled by technological developments outside of Europe. Across the hemisphere, agriculture was developing at a relatively slow (not revolutionary) rate, and Europe over the long run was probably lending and borrowing innovations at about the same rate as several other regions.

Landes believes that the roots of Europe's Industrial Revolution go back to Old Testament times, but things speeded up about a thousand years ago. We are told about the individualism and "institutions of private property" bequeathed to medieval Europeans by the ancient Israelites, Greeks, and Germans (p. 33), about the marvelous freedom enjoyed by medieval European city dwellers and—most important of all—about the unique inventiveness and progressiveness of Europeans. We are told all of these things flatly, without evidence or argument. We are told, more forcefully, about the lack of these advantages and qualities in other contemporaneous societies: China is the main foil. Did the Chinese invent some machine or process during the Middle Ages? If there is any doubt about the matter, Landes criticizes the sinologists for their naiveté. Joseph Needham, for instance, is chided because he focused on Chinese inventions and neglected what Landes considers to be the more important scholarly question: why the Chinese failed thereafter to develop the inventions further. If there is no doubt that the Chinese invented the trait, then Landes either stresses the improvements that Europeans made on it or the unique way that Europeans ("pragmatically") put it to use. And he frames the entire discussion within a global model of China as technologically nonprogressive and, after about 1200, stagnant, even regressive.

Landes then turns to Europe and picks out for display a few important mechanical traits that he thinks were invented in Europe or greatly improved in Europe. He proclaims these traits to be the heart and soul of a medieval industrial revolution. And he dwells on improvements made on the traits much later, long after the Middle Ages ended, thus leaving the reader with the impression that this modern trait was invented and put to use by Europeans far back in the Middle Ages. Thus: an industrial revolution that never happened.

Three technologies are described as touchstones of the supposed medieval industrial revolution, and a word must be said now about each.[37] The three are printing, clocks, and spectacles. Each is proclaimed to have been of great historical significance, both in its historical effects and in its role as a token of the uniquely inventive and innovative European mind. Before the modern research by Needham, Sivin, and others on the history of Chinese technology, it was generally thought that all three of these fields were uniquely European in origin and development. Now we know that this is not the case.

First, as to printing, Landes concedes that the invention of movable metal type was made in East Asia, not Europe, but then he asserts, without evidence, that the Chinese did not really do much reading during or after the Middle Ages (this is of course nonsense).[38] Then he shifts forward to early-modern times in Europe and describes glowingly how many copies of Gutenberg's Bible were circulating in Europe in the sixteenth century, embellishing the argument with comments on the development of printing in England at much later dates. The reader is thus led to believe, falsely, that the intellectual consequences of reading and writing were unique to Europe. I know of no research that shows that Chinese were slower than Europeans to develop printing or to enjoy its benefits.

Clockmaking developed earlier in China than in Europe. The Chinese invented the essential element, the escapement mechanism, more than a thousand years ago.[39] For Landes, however, Chinese clocks and their use was a matter of no historic significance. He describes all of it dismissively and inaccurately.[40] He enshrouds his description in Eurocentric rhetoric: "The Chinese treated time and knowledge of time as a confidential aspect of sovereignty, not to be shared with the people" (p. 50).

(Chinese people paid no attention to time?) "In the cities, drums and other noisemakers signaled the hours" (p. 50). (No sundials, water-clocks, sand-clocks?) The Europeans, in stark contrast, "had to know and order time in order to organize collective activity . . . set a time to wake, to go to work, to open the market, close the market, leave work . . . go to sleep" (p. 50). This describes twentieth-century life, not medieval life.

"The mechanical clock . . . was a European mega-invention of the late 13th century, crucial for its contribution to discipline and productiv-ity" (p. 336). This comment misleads. Clockmaking was not an inven-tion, much less a "mega-invention." It was a complex of small inventions, some in China, some in the Middle East, some in Europe, over a period of several centuries. Indeed, Needham shows that the innovations were crisscrossing the hemisphere throughout this period.[41] Important though the process was, only a polemicist would call it uniquely European. Euro-peans did, however, make an important invention late in the thirteenth century: the clock that is driven by metal weights, not water or sand (though in other ways it was about as mechanical a device as Chinese clocks). We know that it diffused widely and rapidly, but the space-time evolution of Chinese timekeeping devices like the sand-clock is not yet known. Overall, a sober judgment would be: in the fourteenth and fif-teenth centuries Europeans exploited mechanical principles, some neces-sarily borrowed from others, but they did not lead in many areas of tech-nology that were comparable to clocks in historical importance (for instance, the compass, cannons, blast furnaces, printing).[42]

Finally, spectacles. So far as we know, Italians invented them around 1300, but one must take a critical look at Landes's claim that they were a pivotal innovation. Again we are misled, in two ways. First: as with clocks, the role of non-Europeans is neglected, and the path seems to be purely European. But, according to Needham, Chinese were reading with a magnifying glass two centuries earlier.[43] The European invention is a device for two eyes, not one; and Landes treats this innovation as basic: "a wearable device . . . leaving the hand free." He then introduces a strange and unfounded theory about optometry. The lens of the eye "hardens around the age of forty," leading to "farsightedness. . . . [A] me-dieval craftsman could reasonably expect to live and work another twenty years, the best years of his life . . . if he could see well enough. Eyeglasses solved the problem." Landes gives no evidence for this theory, which is sheer speculation about eyes, about craftsmen's work and lives, and more. And he concedes that glasses for nearsightedness did not ap-pear until 150 years later. Spectacles would indeed leave the hands free, and that is important, but just how important? For Landes, the invention of spectacles "more than doubled the working life of skilled craftsmen" (p. 46)—an incredible argument. Eyeglasses "encouraged the invention of fine instruments, indeed pushed Europe in a direction found nowhere else. . . . Europe was already moving toward . . . batch and then mass production" (p. 47). This uncovers the second way that Landes misleads us. He puffs up the importance of this invention (as he does with clocks and printing) to make it seem that medieval Europeans were vastly more inventive than non-Europeans, and that this was both evidence and cause of Europe's supposed advance beyond all other civilizations then and later. Indeed, elsewhere in *The Wealth and Poverty of Nations* he asserts that the roots of the nineteenth-century industrial revolution were well developed in medieval Europe.

Nobody denies that important and consequential inventions were being made in Europe in the Middle Ages. But Landes, like so many other Eurocentric historians in the tradition of Max Weber, draws a picture of medieval Europeans as being uniquely inventive and innovative—uniquely rational—and this is just not credible. The issue is basic. Europeans began to interact intensely with the other civilizations of the Eastern Hemisphere after 1500. If indeed they had the marvelous qualities (and, lest we forget, the marvelous environment) that Landes et al. claim for them, then the later rise of Europe to wealth and power is to be explained by some innate, ancient superiority of Europeans over all others. The alternative theories treat European and non-European civilizations as being, at root, equal in potentiality and comparable in level of development prior to the conquest of the Americas. There are a number of theories then, that account for the rise of Europe relative to other civilizations in terms of facts and forces operating after 1492.[44] One of these is the argument of this book.

EMPIRE

Landes dislikes the empires of Asia, but he likes the European colonial empires. He likes imperialism. He defines it as a sort of natural tendency of societies to expand by conquest. He says imperialism is in our blood: it is "the expression of a deep human drive." "Imperialism has always been with us" (p. 63). Being "a deep human drive," it is not a crass strategy to steal other people's land and riches. "European economies gained little if anything" from it.[45] There have been two really important imperialisms in history: the Islamic expansion and the "expansion of Christian Europe" (p. 393). European expansion started almost as early as the Islamic;

for Landes, European imperialism goes back to the Norse irruption, the Crusades, the *Drang nach Osten*. But the Islamic case, he asserts, was very different from the European case: the Muslim expansion "rested on old ways," was a matter of "zeal," of "passion" (p. 393). European expansion rested on superior technology and was fundamentally an expression of power, profit-seeking, some religious motivation, and curiosity. It was rational. It was also natural. The natives were so backward, ignorant, docile under despotism, that they could not really resist.

Landes's discussion of European imperialism and its postcolonial aftermath takes up about half of the pages of *The Wealth and Poverty of Nations*. It is a key component of his theory of history: that Europe has been superior to everyone else for 1000 years and still remains so in the globalized world of today. But this is much more than a theory of history. It is a political statement; in places even a political diatribe. Landes wants to uphold and defend European-dominated capitalist society in the form that it takes today: he says so over and over again in the book, lashing out at scholars who claim that non-Europeans were and are the equals of Europeans in one way or another; and at scholars who claim to find non-European origins of European cultural advances; and at scholars who criticize Western imperialism as a negative force in history or criticize aspects of Western domination of the Third World today; and at scholars who diminish European greatness by calling imperialism an economic strategy.[46] These critics of Eurocentrism are ideologues; Landes and those who agree with him "prefer truth."

In Chapter 1 of the present book I defined Eurocentric diffusionism as a theory or world model that is grounded in two fundamental arguments; both are essential to its logic. One is the axiomatic argument that Europe naturally progresses, while non-Europe naturally remains backward, traditional, ahistorical. The second is the axiomatic argument that, since 1492, progress for the non-European world has consisted in the diffusion of innovations from Europe. Thus we have a two-sector world in which the European core advances through its own internal forces of progress, whereas the periphery advances mainly by receiving the fruits of Europe's modernization through submission to European domination: or, as Landes puts it, "European (Western) dominion and the fruits therefrom" (p. 63). *The Wealth and Poverty of Nations* follows this logic exactly. Hence the importance of Landes's long, admiring discussion of European imperialism and its present-day outcome.

The history of imperialism is summarized in a thoroughly traditional way. Landes consistently minimizes the negative aspects and exaggerates the benefits. Slavery and slave trading were, he says, merely borrowed from Arab and West African slavers: the structures were in place and Europeans only put slavery to productive use. (In reality, earlier slavery—much of it carried out by Europeans who enslaved Europeans as well as others—was in no way comparable in scale or character to plantation slavery in the European colonies.)

Depopulation of the Americas cannot, of course, be denied by Landes, but he minimizes its importance in the Conquest: we should not dwell so much, in the tradition of Sauer and Borah, on disease as a factor; for Landes it was the technical and cultural superiority of Europeans that really mattered.[47] (Disease was in fact the decisive factor: had New World populations not been decimated by Old World diseases brought over by the Europeans—perhaps three-fourths of the American population died during the sixteenth century—it is likely that the Americans would have quickly picked up European military technology after 1492 and then driven the few thousand European soldiers into the sea.) The Americans, Landes asserts, were cruel, superstitious, in some cases cannibals.[48] The Conquest was natural.

Then we are offered a sequence of false and Eurocentric judgments about plantation colonialism. The slave-plantation colonies were not really important in history. (Absolutely untrue.) The Caribbean climate was too hostile for European settlement. (An environmentalistic myth.) Sugar was of course a very valuable crop, Landes allows, but it did not really have the significance that some scholars (such as Eric Williams) have given it: It did not significantly "alter the path of British development."[49] (This judgment is traditional in European scholarship but has been strongly challenged: see Volume 1.) There was of course resistance by the slaves, but, says Landes, it was partly European-inspired: the Haitian slaves, "encouraged by revolutionary doctrines from France, rose in revolt. . . . The French [were] defeated more by disease than by bullets."[50] (Another traditional myth; in fact, the Haitians made their own revolution. They defeated Napoleon's armies long before Waterloo.) Thus, for Landes, the Europeans emerge as active subjects, the Africans and Native Americans as passive objects.

Basically the same false picture is painted for colonialism in Asia. The British in India merely wanted to engage in peaceful trade; the Indians tried to squeeze them, and this "turned the intruders to thoughts of violence" (p. 154). Pre-British India was supposedly a land of technological backwardness, limited property rights, and great poverty. It was ruled by tyrannical despots. (Not so.[51]) In effect, the British freed the Indians from this unhappy condition by colonizing them. (But colonies, almost by definition, are unfree.) The Dutch in Indonesia would have preferred to be just middlemen, agents, processors, distributors; but conflict with the Iberians back home in essence forced them, against their better judgment, to become colonial rulers. Thus European colonialism, overall, is depicted as something natural in human history, as motivated by all sorts of forces among which the thirst for accumulation was relatively minor, as something that (in spite of its admittedly dark aspects) was beneficial to the natives who, in any case, were unable to do much about it. And decolonization is depicted similarly. In Latin America, "independence slipped in—a surprise to unformed, inchoate entities that had no aim but to change masters." Indonesia was granted independence readily "because [of] Dutch public opinion [and] penitent self-criticism" (p. 149). In the colonies overall, "European ideals of freedom and the rights of man proved contagious, and subject peoples learned from their masters how to resist their masters" (p. 438). In short, everything diffuses from Europe.

Landes also discusses the postcolonial and present conditions in some parts of the Third World, and what emerges is more of the same Eurocentric diffusionist model. The message is: poverty is their own fault, and to progress they must accept diffusions from Europe, most importantly global, Europe-centered, capitalism, in which they are to play a submissive, subaltern role. (This indeed seems to be the crucial message of *The Wealth and Poverty of Nations*. Probably it is the chief reason why *The Wall Street Journal* and the other canonical media gave the book such prominence.) Landes discusses, in turn, postcolonial Latin America, early modern and modern China, the early modern and modern Muslim world, and (very briefly) Africa. He first asks why Latin America, after independence, did not progress as did Anglo-America. There are cultural and environmental reasons, he says. Whereas Anglo-America was settled mainly by English families, Latin America was settled by single males from Iberia who intermarried with the blacks and Indians to form a different sort of society. There is no trace of racism here, but rather a curious judgment that the mixed society of Latin America somehow was bad: "no direction, no identity, no symbolism of nationality, no pressure of expectations. Civil society was absent" (p. 313). He finds another, deeper cause in the fact (as he has it) that Latin America had its origins in Catholic, counter-Reformation Iberia. It was a "simulacrum of Iberian society," lacking "the skills, curiosity, initiatives, and civic interests of North America" (p. 312). Landes of course ranks Europe above non-Europe, but he ranks Catholic Europe much lower than Protestant Europe, and Spain lowest of all. Protestant England was a skeptical, scientific-minded, dissenting, work-ethic society—here a bow to Max Weber—and exported all of this to North America. Catholic Spain was a docile society, ground under "Counter-Reformation orthodoxy and superstitious enthusiasms" (p. 312). (Hardly a balanced judgment.) Latin America's poverty and political instability are mainly explained in this way, but there were also environmental causes: Latin America, being mostly tropical, has a miserable environment for economic development.[52]

Landes scoffs at those who find external factors in the explanation for Latin America's poverty and instabilty. He ignores the fact that only British settler colonies shared in the economic development of modern Britain, that the rest of the world, including Latin America, remained a happy hunting ground for Europeans to exploit. And he acrimoniously denounces the dependency theorists for blaming Latin America's problems on imperialism.[53]

Landes next tells us why (as he believes) China did not develop during the late medieval and early-modern centuries. His argument here is not only traditional Eurocentric history, but it casually ignores a lot of modern scholarly findings. (He cites only a handful of sources, none primary.) The foundation, as we discussed above, is his view of China throughout history as suffering from Oriental despotism and bad reproductive habits, with poverty and nondevelopment as the outcome. But he must contend with the conventionally accepted fact that China was at or above the level of Europe in most aspects of technology until at least the thirteenth century. He concedes that development was taking place before then without explaining how it can be that China's Oriental despotism and Malthusian difficulties permitted this to happen. He then argues, as Eurocentric historians so often do, that something happened in medieval China that stopped technological and economic progress, even led to regression. Landes describes this putative stoppage in colorful terms: "stasis and retreat";[54] "technological oblivion and regression"; "Round, complete, apparently serene, ineffably harmonious, the Celestial Empire purred along . . . " (p. 98); "not only the cessation of improvement but the institutionalization of the stoppage . . . " (p. 200). In a word: China just stopped.

China did not stop. This used to be the conventional view of Western historians. But a great deal of research has been done in recent decades on China's medieval technology, and all of it shows uninterrupted progress overall. Today most historians of China have accepted the hard evidence on this matter. The Eurocentric world historians now retreat to one or another fallback position. Either they focus on some fields of technology in which Chinese progress slowed or stopped—as happens everywhere at one time or another—and ignore or minimize those fields in which progress did not stop; or they zero in on the fifteenth century and claim that some sort of sea change took place in China then and the country backed away from technology.[55] Interestingly, Landes uses these fall-back arguments along with his general assertion of complete stoppage. And, where he concedes some progress, he minimizes its significance. For example, acknowledging that East Asians invented printing with movable type, he says (incorrectly): "Some Chinese printers did use movable type . . . but the technique never caught on as in the West."

In prior chapters I reviewed most of these issues, so I will limit myself here to a few comments.[56] It is wrong to view the history of Chinese technology and economy in terms of the postures taken at various times by the imperial government. Most sectors, including farming, were responding to market forces and local power sources, including landlords.[57] Imperial decrees sometimes favored, sometimes inhibited, the progress in these fields, but these decrees tended to be ignored or easements were purchased with bribes. For example, the traditional view, echoed by Landes, has it that the Chinese backed away from sea trade in the fifteenth century, and indeed imperial decrees were issued at various times forbidding or restricting this trade, but in fact the trade did not stop and perhaps did not even slow down.[58]

The important evidence concerns the sixteenth and seventeenth centuries, not the fourteenth or fifteenth, and it is comparative: European sea trade in Asia increased very rapidly while Chinese did not. But this is to be explained with a completely different model, one that focuses on the dynamics of European expansion, not on a mythical Chinese stoppage. My own theory is set out in Volume 1 and summarized in Chapter 1 of this volume: European expansion overseas resulted from the riches obtained in the sixteenth century in America. Landes to the contrary notwithstanding, Europe had no advantage over Asia, actual or potential, prior to 1492.

Before we leave China, a comment should be made about Landes's mistaken view that Europeans were better at seafaring than Chinese. He has to get around the fact that the greatest maritime accomplishment of the Middle Ages was the series of huge fleets sent out by China in the early fifteenth century, fleets that repeatedly sailed into the Indian Ocean and reached the coast of Africa, all of this some sixty years before Columbus's first voyage. This poses two problems for Landes: First, how does one minimize the significance of this achievement so as to preserve the idea of Europe's absolute superiority in level and rate of technological development? Second, if China reached as far as it did in long-distance maritime voyaging, how do you account for the fact that Europeans, not Chinese, were able to use long-distance voyaging as a stepping-stone to economic modernization and world hegemony? Both problems are dealt with in a standard Eurocentric way. First, the Chinese enterprise itself is subtly diminished: the voyages followed "an orgy of shipbuilding" during which "forests were stripped of timber," masses of illiterate workers were dragooned into building the ships; the ships were "built for luxury"; the enterprise was paid for by "a population bled white by taxes and corvée-levees" (p. 97); the voyages "reeked of extravagance" (p. 514); and so on. And Chinese, he adds (falsely, and with no supporting evidence), were worse navigators than Europeans. This is then contrasted with the more pragmatic, more purposeful, more rational voyaging by the Europeans. But Landes here is merely using literary devices to disguise (and distort) what was certainly one of the most important technical accomplishments of the Middle Ages.

As to the matter of long-range implications for world history, Landes falls back on a string of familiar arguments, all of which have been refuted by modern scholarship, which he dismisses: "We have seen examples of this Europhobia in recent discussions of the age of voyages and discovery" (p. 96). The Chinese, in essence, did not have the rationality to build on the base of Admiral Zheng He's voyages: the Chinese "lacked range, focus, and above all, curiosity" (p. 96). And their society was stagnating, regressing: in essence, it was just Oriental despotism. But China was neither stagnating nor regressing. The great voyages served their purpose, and (after about 1440) the government found it more rational to devote its resources to defending the land frontier in the northwest rather than to underwrite further voyages. Landes asks, Why didn't the Chinese voyages reach America? and answers with all of the arguments just discussed about stagnation, lack of curiosity, and so on. But the answer, as we saw in prior chapters, is straightforward. The Americas are much, much closer to Europe than to China. And Europe had a strong reason to venture out into the Atlantic: the hope of reaching the wealth of Asia. The Chinese had no such motive: what possible advantage would have accrued to China if it had found a direct sea route to Europe? Chinese merchant ships continued to sail to and from Southeast Asia and other foreign destinations; there was no interruption. As in other aspects of technology and society, Chinese were as progressive as Europeans—and as rational.

Landes next turns his attention to the Middle East. He claims that the Islamic world, including Mughal India, simply had no chance to rise as Europe did. Its history is "history gone wrong."[59] His discussion of the medieval and early-modern Islamic world is fundamentally erroneous. He falsifies some aspects of early Islamic society. And he claims falsely that

Islamic society, throughout its history, is characterized and defined by selected unmodern features of some versions of the Islamic religion, and selected unmodern and undemocratic contemporary Islamic societies. (It is
rather like defining European history as a combination of the Inquisition
and Nazi Germany.) Says Landes about the religion and the society (including here Mughal India): "Islam links faith to power and dominion. . . . Europe was spared the thought control that proved a curse to Islam" (p. 394). "Unlike Islam . . . Christianity early made the distinction
between God and Caesar" (p. 38). "Islam does not, as Christianity does,
separate the religious from the secular. . . . The ideal state would be a the-
ocracy. . . . A good ruler leaves matters of the [mind] to the doctors of
faith. This can be hard on scientists" (p. 54). To which one must reply:
the Islamic societies of the Middle Ages had as much freedom of thought,
as much science, as did the Christian world. This is widely accepted by
scholars of medieval Islam.[60] As to modern Islam, among many uncom-
plimentary assertions by Landes, this one stands out: "[The] ill is far more
general than the Israeli-Arab conflict. . . . It lies . . . with the culture"
(p. 410). I see no need to comment.

About Africa, little is said, apart from a few assertions about the
primitiveness of the cultures and the inhospitability of the climate. Here
are a few of Landes's comments, taken out of context but conveying his
tone: "Sub-Saharan Africa threatens all who live or go there." "Traditional nostrums and magical invocations may be preferred [by Africans]
to foreign, godless medicine" (p. 12). "The slave trade flourished" in
precolonial Africa (p. 69). Small farmers "scratch the soil" (p. 500). "In
general, the women do as they are told. . . . AIDS? Forget condoms; the
men don't like them" (p. 501). And a general characterization of
postcolonial sub-Saharan Africa: "bad government, unexpected sovereignty, backward technology, inadequate education, incompetent if not
dishonest advice, poverty, hunger, disease, overpopulation" (p. 499). If
these negative judgments were backed with scholarly evidence they
might deserve to be taken more seriously. They reflect ignorance.

A SUMMING-UP

Landes says, at the beginning of *The Wealth and Poverty of Nations*, that
he prefers truth to ideology. He believes that what he is telling us is truth,
and I am sure that he is honest in this belief. But most ideologues believe
that what they pronounce is the truth. Landes presents only that part of
the truth about world history that makes Europeans look good and non-

Europeans look bad. He relates, as truth, a mass of old myths that tell of Europe's past and present superiority and priority. The modern scholarship that questions these myths is either denounced as heretical or (much more frequently) ignored.

This is ideology, not truth. I suggest, therefore, that we view this book as a recounting of Europe's myths of origin, not as a work of historical scholarship, which it is not.

NOTES

1. David Landes, *The Wealth and Poverty of Nations: Why Some Are So Rich and Some So Poor* (1998). Page numbers in parentheses in the text refer to this work.

2. As we noted in Chapter 1, two other foundation theories were popular in the past: one grounded in religion, the other in race. See Volume 1 for further discussion.

3. See Peet, *Modern Geographical Thought* (1998).

4. See, for example, Huntington, *Civilization and Climate* (1924). Many of Landes's environmentalistic arguments are borrowed from Eric L. Jones's *The European Miracle* (1981), which was discussed in Chapter 5. Recall that we are using "environmentalism" and "environmental determinism" as synonyms.

5. See, for example, Collins and Roberts, *Capacity for Work in the Tropics* (1988). According to Landes, "damp, 'sweaty' climes reduce the cooling effect of perspiration" (p. 6). This is absurd: perspiration is precisely a response to ambient heat.

6. "It is no accident that slave labor has historically been associated with tropical and semitropical climes." Then Landes has a footnote approvingly quoting Adam Smith: "The constitution of those who have been born in the temperate climate of Europe could not, it is supposed, support the labor of digging the ground under the burning sun" (p. 7). Then Landes again (p. 9): "The solution [to the climate problem] was found in slavery."

7. Biomass production potential is highest in the humid tropics and subtropics; some tropical crops (sugar cane and rice, for instance) can give higher yields than any midlatitude crops, all other things being equal (fertilization, soil moisture relations, and so on).

8. The World Health Organization estimates that tropical diseases, including malaria, kill about one-quarter as many humans per year as do respiratory diseases, most of which are not as important in the humid tropics as in cooler regions. See Porter and Sheppard, *A World of Difference: Society, Nature, Development* (1998), pp. 211-259; and see Chapter 8.

9. See Giblin, "Trypanosomiasis Control in African History: An Evaded Issue?" (1990); Turshen, "Population Growth and the Deterioration of Health: Mainland Tanzania, 1920–1960" (1987); and Volume 1. Both humans and cattle had, and have, partial immunity to trypanosomiasis in Africa; see Blench (1993).

10. On this matter see Volume 1. Another false assertion by Landes: there are "incredibly high rates of evaporation" in the tropics (p. 14). Evaporation and evapotranspiration rates in the humid tropics are roughly comparable to those in humid midlatitudes during the summer. Year-round, they are about double those of the latter, but not "incredibly high." Still another falsehood: "towns cannot thrive in tropical Af-

rica" because of poor food supply, resulting, supposedly, from the limitations of tropical agriculture (p. 13). Nonsense: look at all the great cities of Africa. And still another falsehood, in fact two in one: "Monsoon rains . . . vary a lot from season to season and year to year. Floods and droughts are the norm" (p. 28). Rainfall variability in monsoon regions, notably India and Pakistan, is serious in semi-arid areas, as elsewhere. Floods and drought, also, are no worse here than in many other regions, although their effects are worse—because of poverty.

11. Landes also makes a few false assertions about arid climates and uses these as the groundwork for his theory about the unprogressive, despotic nature of the Middle East. We return to this subject later.

12. In much of the North European Plain, agriculture did not become productive until the moisture-loving potato arrived from South America, its place of domestication.

13. Says Landes (falsely): European horses were better than all others, hence the "battle steeds" were superior. Europeans had "an advantage in heavy work and transport." Horse manure meant higher fertility of soil, and so "Europeans kept a diet rich in dairy products, meat, and animal proteins" (all on p. 20). This is old-fashioned Euro-environmentalistic nonsense. See Chapter 3.

14. The title of Chapter 2 of The Wealth and Poverty of Nations is "Answers to Geography: Europe and China."

15. Although The Wealth and Poverty of Nations is written for a popular readership, Landes has a scholar's responsibility, which he shirks, to inform his readers of differences of scholarly opinion. Most statements in this book are asserted as plain fact, even though many of them are highly controversial.

16. Oriental despotism is also alluded to on pp. 27–28, 34, 39, 111, 156–58, 313, and 410. See Chapter 2, note 7, of this volume.

17. See Chapter 5 and Volume 1, pp. 80–92; also, Blaut, "Where Was Capitalism Born?" (1976).

18. "The concept of rights went back to Biblical times and was transmitted and transformed by Christian teachings" (p. 34).

19. "The Hebrew hostility to autocracy . . . set the Israelites apart from any of the kingdoms around"—these latter being, of course, the Asian "despotisms" (p. 34).

20. "Ironically . . . Europe's great good fortune lay in the fall of Rome and the weakness and division that ensued" (p. 37). "[Property] rights had to be rediscovered and reasserted after the fall of Rome" (p. 33). Michael Mann holds the same view (see Chapter 6).

21. Landes: "In China, even when the state did not take, it oversaw, regulated, and repressed. Authority should not have to depend on goodwill, the right attitude, personal virtue" (p. 35). This is nonsense. See Hucker, "Ming government" (1998). Also, see Chapter 8 and Volume 1, pp. 107–108. On oligarchic republics of ancient India, comparable in some ways to ancient Athens, see Mukerji, The Republican Trend in Ancient India (1969).

22. See Volume 1, pp. 131–132.

23. Landes: "Fragmentation gave rise to competition." (p. 36) Some Eurocentric historians claim that the brotherhood of feudal aristocrats was itself a democracy and devolved later into democratic European states. See my comment on this theory in Chapter 6.

24. See Hucker, "Ming government," p. 105; Heijdra, "The Socioeconomic Development of Rural China During the Ming" (1998), p. 571; Goody, The East in the West (1996), pp. 230–231.

25. Goody, *The East in the West*; Frank, *ReORIENT* (1998); Rowe, *Hankow: Commerce and Society in a Chinese City, 1769–1889* (1984); Champakalakshmi, *Trade, Ideology and Urbanization: South India 300 B.C. to A.D. 1300* (1996).

26. Chapter 3 in Landes's book is titled "The Invention of Invention."

27. See, for example, Needham, *Science and Civilization in China* (1954–); Kuppuram and Kumudamani, *History of Science and Technology in India* (1990); al Hassan and Hill, *Islamic Science Technology* (1986); Watson, *Agricultural Innovation in the Early Islamic World* (1983).

28. Page 38. Also see pp. 24, 35, 27, 98; for India, pp. 14, 157, 395. See Ross (1998) on Malthusianism.

29. Page 22. Other Malthusian comments: pp. 21, 23, 24, 187, 345, and 499.

30. See Goody, *The East in the West*, pp. 138–161, and references in notes 11 and 12 of Chapter 5 of this volume.

31. Page 40. This revolution is described in the chapter titled "The Invention of Invention" (pp. 45–59).

32. This theory, positing a sort of take-off by Europe roughly one thousand years ago, is very widely held among present-day Eurocentric historians, including Lynn White, Jr., Eric Jones, and John Hall (see Chapters 3, 5, and 7 of this volume), all of whom are abundantly cited by Landes, and Michael Mann (Chapter 6), who is not cited but develops the theory most fully.

33. Page 54. "History Gone Wrong" is the title of a chapter that denounces Islamic society—I cannot phrase it more gently—throughout history.

34. As we saw in earlier chapters, this error was often made by Eric Jones, John Hall, and Robert Brenner. All but the last are cited by Landes as authority, and so the errors reproduce themselves.

35. Page 220. In China, peasants were "human cattle" (p. 37) and there was "absence of freedom" (p. 56). This is of course nonsense.

36. See, for example, Grantham, "Contra Ricardo: On the Macroeconomics of Pre-Industrial Economies" (1999); Titow, *English Rural Society, 1200–1350* (1969); Clark and Van Der Weif, "Work in Progress: The Industrious Revolution" (1998); Smith, *An Historical Geography of Western Europe* (1967).

37. Two other technologies also are emphasized: gunpowder and the water wheel. These are discussed elsewhere in this chapter.

38. Another Western conceit: "ideographic writing works against literacy" (p. 51).

39. I am indebted to Nathan Sivin for pointing out to me the crucial significance of the Chinese invention of the escapement mechanism. See, in Needham, *Science and Civilization in China: Vol. 4, Part 2: Physics and Physical Technology: Mechanical Engineering* (1965), pp. 435–545, the extensive discussion of the evolution of clocks in Europe and the Middle East as well as China. See Sivin, "Why the Scientific Revolution Did Not Take Place in China—Or Didn't It?" (1984).

40. Inaccuracies: "Chinese never got beyond water-clocks" (p. 50). See Needham, *Science and Civilization*, Vol. 4, Part 2, on the more advanced and all-weather Chinese sand-clocks. "Chinese built a few astronomical water-clocks in the Tang and Sung eras. . . . These monumental machines were imperial projects" (p. 50). This misleads a reader because it refers to very early times—later, during the Ming, sand-clocks were widely distributed. See Needham, loc. cit.

41. Needham, *Science and Civilization in China. Vol. 4, Part 2*, pp. 532–546. See especially the diffusion diagram on p. 533.

42. Needham, *Science and Civilization in China. Vol. 4, Part 2*, pp. 544–545. Also see

al Hassan and Hill, *Islamic Technology*; Kuppuram and Kumudamani, *History of Science and Technology in India* (1990).

43. Needham, *Science and Civilization in China. Vol. 4, Part 2*, p. 120.

44. We should place along with these theories the formulation of Janet Abu-Lughod: the hemisphere was essentially a system of interconnected societies, in which Europe's level of technology was somewhat lower than Asia's until about 1350; then, she believes, the Black Death struck Asia much harder than it did Europe, and this led to the initiation of the relative rise of Europe. Note that this theory is completely non-Eurocentric, although it places the origins earlier than I do (1492).

45. Page 429. Also see p. 423.

46. See pp. 4–5, 63, 103–104, 107, 119–121, 163, 165, 216–218, 225, 321, 326–328, 346–349, 409, 415–418, 423–425, 432, 438, 525, 551, 553, 557, 565.

47. Carl Sauer, the foremost American geographer of his time, is identified by Landes as an "agricultural anthropologist-archaeologist" (p. 25). Obviously Landes is ignorant about the science of geography. This would be a trivial matter were it not for the emphasis that Landes places in the first two chapters of his book on what he thinks is scientific geography (but is not that), and for his comment at the beginning of Chapter 1 deploring the low status of geography in the American academy. We do need to improve the status of geography, but not the environmentalistic geography propounded by Landes. I write as a geographer.

48. The Aztecs are of course a good target of opportunity; Landes describes one of their noblemen as "the Mexican Darth Vader"; as a "prince of darkness" (pp. 104–105). Nowhere are any Europeans so described.

49. "Third World countries and their sympathizers want to enhance the bill of charges against the rich, imperialist countries" (p. 122). Eric Williams is mischaracterized as a Marxist: "he reduces everything to economic motives and interests" (p. 119).

50. Page 117. On slave ships "it was hard to deal kindly, if only because a slave ship's atmosphere reeked of fear and hate" (p. 118). Sympathy for the slavers?

51. See Richards, "Early Modern India in World History" (1997); Habib, "Merchant Communities in Precolonial India" (1990); Kuppuram and Kumudamanik, *History of Science and Technology in India*; Perlin, *The Invisible City* (1993); Subramanian, *India's International Economy, 1500–1800* (1999); Frank, *ReORIENT*; Goody, *The East in the West*.

52. Landes comments that Argentina is not burdened with a tropical environment, but all of the social reasons just discussed explain the underdevelopment of Argentina.

53. One of Landes's comments on dependency theories: "By fostering a morbid propensity to find fault with everyone but oneself . . . [dependency doctrines] promote economic impotence." Then, in italics: *"Even if they were true, it would be better to stow them"* (p. 328). One senses here a bit of ideology.

54. The title of one chapter is "The Celestial Empire: Stasis and Retreat" (pp. 335–349).

55. It is often argued that the fifteenth century in Europe was also a time of technological stasis, or at most very slow progress. See Lopez and Miskimin, "The Economic Depression of the Renaissance" (1961–1962); Thorndyke, "Renaissance or Prenaissance?" (1943); and Chapter 2 of Volume 1.

56. Also see Volume 1, Chapters 2 and 4.

57. Pomeranz, *The Making of a Hinterland: State, Society, and Economy in Inland North China 1853–1937* (1993); Wong, *China Transformed* (1997); Marks, *Tigers, Rice,*

Silk, and Silt (1998); Goody, *The East in the West*; S. Mann, *Local Merchants and the Chinese Bureaucracy, 1750–1950* (1987).

58. Heijdra, "The Socioeconomic Development of Rural China During the Ming" (1998); Wang, "Merchants Without Empire: The Hokkien Sojourning Communities" (1991); Volume 1, p. 181.

59. "History Gone Wrong" is the title of the chapter in *The Wealth and Poverty of Nations* that deals mainly with the Islamic world (pp. 392–421).

60. See, for instance, Hodgson, *The Venture of Islam* (1974); al Hassan and Hill, *Islamic Technology* (1986); Rodinson, *Islam and Capitalism* (1973); Watson, *Agricultural Innovation in the Early Islamic World* (1983).

CHAPTER 10

Thirty Reasons Why Europeans Are Better Than Everyone Else (A Checklist)

B y my count, thirty different reasons have been put forward by our eight Eurocentric historians to explain Europe's supposed superiority or priority in ancient, medieval, and early-modern times. I suspect that these thirty propositions include most of the Eurocentric arguments that are used in this manner by historians today. Recall that an argument is described as "Eurocentric" in this book when it *falsely* favors Europe or Europeans over other peoples and other places.

In this chapter I will provide a checklist of the thirty reasons, mainly to demonstrate the breadth of Eurocentric arguments in history. In the following chapter I will show how these arguments are woven together (leaving a few loose strands) into a general model that is, with some qualifications, put forward by seven of the historians discussed here.

The checklist of arguments will be presented as a numbered series of propositions, and, for each, the historians who put forward that argument will be named.[1] No comments will be offered.

1. People of the white race have an inherited superiority over the people of other races. (Weber argued this way; but none of the seven contemporary historians expresses racist views.)

2. The climate of Europe, or northwest Europe, is uniquely favorable for agriculture. (Jones, Mann, Hall, Landes) Or: Europe, along with China, possesses a climate that is more favorable for agriculture than are the climates of all other regions, especially the humid tropics. (Diamond)

3. The climate of Europe is better for human comfort and productivity than are the climates of all other regions. (Jones, Landes)

4. The soils of Europe are uniquely fertile. (Jones, Mann, Hall, Landes)

5. The landform structure of Europe is uniquely favorable for communication and the diffusion of ideas. (Jones, Diamond, Landes)

6. The landforms of Europe differentiate the continent into separate ecological cores, and this explains in large part the fact that Europe has many moderate-sized states instead of an empire. (Jones, Hall, Diamond, Landes)

7. The indented coastline of Europe partly explains the linguistic, ethnic, and political differentiation of Europe. (Jones, Mann, Diamond)

8. The forest vegetation of Europe historically contributed to the development of individualistic people and small families, hence led Europe toward private property and capitalism (Weber, Mann, Hall, Landes) and helped Europe uniquely to avoid overpopulation and Malthusian disasters. (Mann, Hall, Landes)

9. Europe's environment is less subject to natural disasters than are other regions, and this encouraged development. (Jones, Hall)

10. Europe was, historically, less disease-ridden than all other places. (Jones, Diamond, Landes)

11. Europeans, historically, were better nourished than other people. (White, Jones, Landes)

12. Europeans were uniquely inventive. (Weber, White, Brenner, Jones, Mann, Hall, Landes)

13. Europeans were uniquely rational in the practice of sexual self-restraint and so avoided overpopulation and Malthusian disasters. (Jones, Hall, Landes)

14. Europeans were uniquely innovative and progressive. (Weber, White, Brenner, Jones, Mann, Hall, Diamond, Landes)

15. Europeans were uniquely capable of creative and scientific thought. (Weber, White, Mann, Hall, Landes)

16. Europeans held uniquely democratic, ethical values. (Weber, White, Mann, Hall, Landes)

17. The development of classes and/or class struggle was most fully developed in Europe. (Weber, Brenner, Mann, Hall, Landes)
18. The Christian religion, as doctrine, led to unique European development. (Weber, White, Mann, Hall)
19. The Christian Church, as institution, led to unique European development. (Weber, White, Mann, Hall, Landes)
20. The European family was uniquely suited to development. (Also see No. 8.) (Jones, Mann, Hall, Landes)
21. Europeans uniquely, in ancient and/or medieval times, developed the concept and institution of private property. (Weber, White, Brenner, Jones, Mann, Hall, Diamond, Landes)
22. Europeans uniquely, in ancient and/or medieval times, developed the institution of the market. (Jones, Hall, Diamond, Landes)
23. Urbanization, in Europe, was more favorable for development than elsewhere; European cities were more progressive and/or more free than cities elsewhere. (Weber, Jones, Hall, Diamond, Landes)
24. The state, in Europe, developed toward modern politics more rapidly than elsewhere. (Also see Nos. 25, and 26) (Weber, Jones, Mann, Hall, Diamond, Landes)
25. The empire as a political form hobbled development in non-European regions. (Weber, Jones, Mann, Hall, Diamond, Landes)
26. Oriental despotism hobbled social and technological development in non-European regions. (Also see No. 25) (Weber, Jones, Mann, Hall, Diamond, Landes)
27. Europe was uniquely capable of avoiding Malthusian disasters for many reasons. (Also see Nos. 8 and 13) (Brenner, Jones, Mann, Hall, Landes)
28. The practice of, and dependence on, irrigation slowed or stopped development in hydraulic or irrigating societies. (Also see No. 26) (Weber, Jones, Mann, Hall, Landes)
29. The development of feudalism in Europe uniquely favored the rise of democracy and private property. (Also see No. 21) (Weber, Jones, Mann, Landes)
30. Europeans were uniquely venturesome, uniquely given to exploration and overseas expansion. (Jones, Mann, Landes)

The reader may note that religious faith, which, as I mentioned in Chapter 1, was one of the four prime categories of Eurocentric historical explanation in earlier times, does not appear on this checklist. Race, likewise, appears only as a view held by Weber almost a century ago. Neither of these views is expressed by the seven contemporary historians.[2] The reader should note also that additional arguments would have to be added if we were discussing modern history: for instance, Eurocentric explanations for the Industrial Revolution.[3] And, of course, a larger checklist would be needed to catalog the dreary list of negative arguments about specific parts of the non-European world, arguments amply discussed in prior chapters. (For instance, Chinese preferred "copulation above commodities," according to Jones; India "had no sense of brotherhood," according to Hall; African women "do as they are told," according to Landes.) Finally, the reader should note that my division of the corpus of Eurocentric history into these thirty segments is somewhat arbitrary. There is one entire cake, but it can be sliced in various ways.

We can of course try to count the number of arguments used by each historian and the number of historians who employ each argument, but this exercise has somewhat limited value. This is true partly because the cake has, indeed, been sliced up somewhat arbitrarily; partly because different historians emphasize different arguments and the fact that one or another neglects a certain argument doesn't tell us that he (all eight are males) doesn't accept this argument as valid; and partly because the various scholars disagree somewhat as to which arguments are truly fundamental, although with one exception (Brenner, the Marxist), the modern historians largely share a common explanatory model, which we discuss in Chapter 11. Neglecting these qualifications, however, we can see a few interesting numbers.

Five of the scholars, Jones, Mann, Hall, Diamond, and Landes, seem to want to relate to us *all* of the important reasons for Europe's superiority throughout history. Diamond's overwhelming emphasis on environmentalistic arguments (what he calls the "ultimate" reasons for European superiority) obscures this fact, but he also puts forward five of the sociocultural arguments (as "proximate" reasons). It seems best to set Diamond aside and to consider the commonalities among Jones, Mann, Hall, and Landes.

Twelve of the thirty arguments are used by all four of these historians. Obviously there is a degree of consensus here, and we will discuss the meaning of this partial consensus in the next chapter. Landes stands out: he employs twenty-five of the thirty arguments. Clearly he wants to throw into the stewpot of European superiority virtually all of the ingredients at hand.[4]

Not much can be learned by counting the number of historians who utilize each of the thirty arguments. The relatively low values for environmentalistic arguments (soil, climate, and so on) probably reflect the fact that most of the scholars are social scientists, but the pattern does suggest that environmentalistic arguments are indeed favored to some extent by all of them: the favorite argument is the superiority of Europe's environment for agriculture. Landes, Diamond, and Jones are particularly prone to use the arguments of environmental determinism. The Malthusian arguments are also among the favorites. But the over-whelmingly important arguments are Weberian, invoking claims about the superiority of the European mind: its rationality, inventiveness, inno-vativeness, venturesomeness, and so on. Max Weber is still the godfather of Eurocentric historiography.

NOTES

1. Where a historian is named in connection with a specific argument, this, in al-most all cases, refers to matters discussed in prior chapters. In a few cases, mention of the name indicates that the historian has asserted the argument in other writings (this is es-pecially true for Weber). For Jones, I use the arguments asserted in *The European Miracle* and not actually retracted in his later book *Growth Recurring*. For most propositions, I do not indicate the historical period that is referred to by a given historian: some assert the proposition for ancient times and thence forward; some for medieval and early-modern times; some for all of history.

2. Some of the arguments made by Lynn White, Jr., in his book *Machina Ex Deo: Essays in the Dynamism of Western Culture* (1982) seem (to me) to be saying that his views about the rise of the West were partly grounded in, or perhaps partly confirmed by, his own religious faith (see Chapters 3, 4, and 5 in that book). K. F. Werner, a German medi-evalist whose views were discussed in Volume 1, pp. 147–148, seems also to be telling us that his historical views are influenced by his faith: see his 1988 essay "Political and So-cial Structures of the West, 300–1300." A historian is eminently justified in stating the underlying world view that informs his or her work. It would be interesting to find out whether the earlier view that God has helped Europeans to advance to civilization, and to diffuse that civilization over the earth, is held, unconsciously, by other contemporary historians.

3. I plan to suggest a non-Eurocentric explanation for the Industrial Revolution in Volume 3 of *The Colonizer's Model of the World*.

4. "Landes reasserts in his new book the dominant Eurocentric interpretations of modern history that have prevailed among European and North American historians for at least two centuries" (Buck, "Was It Pluck or Luck That Made the West Grow Rich?", 1999, p. 417). Also see Goldstone (2000).

CHAPTER 11

The Model

The Eurocentric historians chosen for analysis in this book were chosen precisely because they *are* Eurocentric, and they do not reflect the views held by historians in general. But most Western historians seem to share some of the Eurocentric views that we have been discussing. Stating the matter differently: there seems to be a model of world history that is widely accepted today, and it is not free of Eurocentric errors.

Some historians argue that Europe (or the West) was more advanced, more developed, than all other societies in 1500. Many historians, however, accept the newer evidence (particularly about China) that other societies were at a level as high as or higher than Europe in 1500 in terms of technological and economic development. But most of these latter historians argue, in a way that I view as Eurocentric, that Europe in 1500 had a unique *potential* for development: usually a matter of Weberian rationality in such things as inventiveness, progressiveness, venturesomeness, and the like.

Therefore most historians appear to accept the following proposition: the uniqueness of Europe was already in place by 1500; that is, the rise of Europe, in absolute and relative terms, had already begun before the discovery of America. Explanations then refer to processes that are thought to have been at work in ancient times, or medieval times, or both, or during an even longer period (as Weber, Diamond, Jones, Mann, and Landes argue). For both groups of scholars, those who believe that Europe was uniquely advanced by A.D. 1500, and those who believe that Europe had a unique potential in 1500 to advance later—to rise, modernize, and so forth—the basic problem is to explain *why* Europe was more advanced or more capable of future advancement than all other societies at the beginning of the modern era. On this question there are of course many points of view, ranging from the Eurocentric Marxism of scholars like Robert Brenner to the conservative Eurocentrism of scholars like Jones and Landes. Perhaps it can be said of all Eurocentric historians that they agree on the outcome but disagree (often with great vehemence) on the reasons why that outcome came to be.

Six of the eight historians discussed in this book accept some tenets of environmental determinism as either a minor cause, or an important cause, or the most important cause (Diamond) for Europe's historical superiority (or priority). Indeed, most of the recent works on world history that I have seen, including textbooks, seem to incorporate at least a bit of environmentalism in their explanations for the relative and absolute rise of Europe. There seems to be wide acceptance of the old theory (expressed, as we saw, by Weber, but much older in point of origin) that Europe's climate gave it specific advantages over arid and tropical regions (including India). Arid regions supposedly require irrigated agriculture, and, for many historians, irrigation had various awful consequences, notably permanent despotism. Peoples of the humid tropics (not always excluding India) are dismissed as historical nonactors: explicitly so by a few historians including of course Jones, Diamond, and Landes; implicitly so by many others. The other face of this environmentalistic view is a set of propositions about Europe's supposedly superior environment: abundant rainfall, good soils, topographic differentiation into "cores" that encourage ethnic, economic, and political diversity and immunize Europe against empire, and so on. At least a few of these arguments (nine of which were mentioned in our checklist) are used, as far as I have been able to tell, by most other Western historians as part of their explanations for the rise of Europe.

The second cause of Europe's putative superiority is culture, but for many historians, including the eight discussed in this book, this is reduced to rationality: to a mentality that favors invention, innovation, and the rest. Although Jones (in *Growth Recurring*) criticizes Weber, he, along with all the other historians discussed in this book, describes Europeans as more inventive (and so on) than everyone else, and ultimately expresses a Weberian view. Brenner discovers unique European rationality as a product of the late-medieval rise of capitalism, which supposedly brought with it a unique (Weberian) rationality in invention and the like. Diamond incorporates rationality into his explanation for the rise of Europe above China. Thus the essentially Weberian theory that there existed a unique European (or Western) rationality, and this contributed strongly to the rise of the West, is accepted by all eight of these scholars. It is accepted by many other historians today.

Most of the historians discussed here also accept the very old theory that Europe's culture has two roots: the supposedly individualistic, aggressive virtues of the northern European tribes, and the intellectual, moral, and religious contribution of the Mediterranean peoples, the ancient Greeks, the Romans (Mann and Landes apparently demurring[1]), and thereafter early Christianity and the Western church. All of them focus attention on the rise of a kind of capitalist personality, private property, and free markets, as a major causal force, but all the arguments claiming European uniqueness in these matters seem to me to be grounded in the notion that these are products of European rationality, which in turns goes back to the two primal sources, Indo-European tribal qualities and Graeco-Judeo-Christian inheritance. Many other historians accept this twin-root theory of the origins of European uniqueness. It is an old theory: you find it, for instance, in both Marx and Weber.

The same argument-form seems to underlie assertions about European superiority (or precocity) in the invention or development of social institutions. The theory of a uniquely progressive western-European family, formed into an essentially Malthusian explanation for Europe's supposedly unique ability to keep population in check, for the early individualism and entrepreneurship of Europeans, and so on, is much emphasized by Eurocentric historians at present; among the seven modern historians whom we are discussing, the theory is explicitly asserted—even though it is not emphasized—by Jones, Mann, Hall, and Landes.[2] Interestingly, only Jones, in his later work *Growth Recurring*, mentions the fact that newer scholarship strongly questions this theory: nuclear families were not unique to medieval Europe and seem to have been an effect, not a cause, of modernization.[3] The explanation for the supposed uniqueness of the Western family, again, finds it to have its main origins in the supposed unique individualism of the ancient Indo-Europeans. Similarly for other institutions, like markets, cities, and so on, the social fact tends to be rooted in some primordial mental fact: essentially, Weberian rationality.

But the strongest arguments for Greater Europe's ancient and medieval superiority are negative ones: assertions, sometimes very harsh ones (especially in *The European Miracle* and *The Wealth and Poverty of Nations*), about the *inferiority* of mentalities, societies, and environments outside of Europe. Four of the historians discussed here (Jones, Mann,

Hall, and Landes) make such assertions about non-Europeans. Most other historians do not do so.

An extension of these views, and thus a part of the Eurocentric model, is the notion, expounded vigorously by Landes and also by Brenner, that non-European portions of the world not only were relatively unimportant for social evolution before 1500 but remained so during the age of colonialism and are still basically unimportant today. The dynamics of Europe have always been within Europe and remain so. (Parts of East Asia have recently been elevated to the status of honorary European.) Thus a historical view becomes a political one. Europe, some historians say, has ruled the world—benignly: diffusing civilization—for five hundred years, and so it is natural, and *right*, for Europe to dominate the world today and tomorrow.

There are differences, of course, in the viewpoints expressed by the various historians. Some stress religion as a causal factor; others demur. Some stress the natural environment; others accept it as significant but consider it to have been a minor factor. Some stress economics; others stress politics. And so on.

But most historians seem to accept the general model: Europe, before modern times, rose above all other societies because of its uniquely progressive mentality and its uniquely bountiful environment. Europe was somehow the natural center of the world.

I disagree.

NOTES

1. Landes: "Europe's great good fortune lay in the fall of Rome and the weakness and division that ensued"; in *The Wealth and Poverty of Nations* (1998), p. 37. Also see Mann, "European Development: Approaching a Historical Explanation" (1988), p. 16.

2. I discussed the theory of the unique European family—unique, supposedly, in sexual restraint, late age of marriage, low marriage rate, tendency toward the nuclear rather than the extended household, and more—in Volume 1 (pp. 66–68 and 128–135). The historians whom we are discussing say little about this theory but cite most of the important proponents of the theory, among them Laslett ("The European Family and Early Industrialization," 1988); Crone (*Pre-Industrial Societies*, 1989); Stone (*The Family, Sex and Marriage in England 1500–1800*, 1977); and Macfarlane (*The Origins of English Individualism*, 1978).

3. Jones firmly supported the theory in *The European Miracle* (1981).

Bibliography

Abu-Lughod, J. 1989. *Before European Hegemony: The World System A.D. 1250–1350.* New York: Oxford University Press.

Abun Nasr, J. M. 1975. *A History of the Maghrib.* 2nd ed. Cambridge: Cambridge University Press.

al Hassan, A., and Hill, D. 1986. *Islamic Technology.* Paris: UNESCO.

Amin, S. 1974. *Accumulation on a World Scale.* New York: Monthly Review Press.

_____. 1989. *Eurocentrism.* New York: Monthly Review Press.

Anderson, P. 1974. *Lineages of the Absolute State.* London: New Left Books.

Appadorai, A. 1936. *Economic Conditions in Southern India (1000–1500 A.D.).* 2 vols. Madras: University of Madras Press.

Arasaratnam, S. 1996. *Maritime India in the Seventeenth Century.* Delhi: Oxford University Press.

Aston, T. ,and Philpin, C., eds. 1985. *The Brenner Debate: Agrarian Class Structure and Economic Development in Pre-Industrial Europe.* Cambridge: Cambridge University Press.

Baechler, J. 1988. The Origins of Modernity: Caste and Feudality (India, Europe and Japan). In Baechler et al., eds., pp. 39–65.

_____, Mann, M., and Hall, J. A., eds. 1988. *Europe and the Rise of Capitalism.* Oxford: Blackwell.

Barrett, W. 1991. World Bullion Flows, 1450–1800. In Tracy, ed., pp. 224–254.

Beckett, J. 1990. *The Agricultural Revolution.* Oxford: Blackwell.

Bennett, M. 1954. *The World's Food: A Study of the Interrelations of World Populations, National Diets, and Food Potentials.* New York: Harpers.

Berkner, L. 1975. The Use and Misuse of Census Data for the Historical Analysis of Family Structures. *Journal of Interdisciplinary History* 5:721–738.

Bernal, M. 1987. *Black Athena: The Afroasiatic Roots of Classical Civilization: Vol. 1. The Fabrication of Ancient Greece.* London: Free Association Books.

Black, C. 1966. *The Dynamics of Modernization: A Study in Comparative History.* New York: Harper & Row.

Blaut, J. 1958. Chinese Market-Gardening in Singapore. Ph.D. dissertation, Louisiana State University.

_____. 1962. The Nature and Effects of Shifting Agriculture. In *Symposium on the Impact of Man on Humid-Tropics Vegetation*. Canberra: UNESCO and Australian Government Printer.

_____. 1963. The Ecology of Tropical Farming Systems. *Revista Geográfica* 28:47–67.

_____. 1976. Where Was Capitalism Born? *Antipode* 8(2):1–11.

_____. 1987a. *The National Question: Decolonizing the Theory of Nationalism*. London: Zed Books.

_____. 1987b. Diffusionism: A Uniformitarian Critique. *Annals of the Association of American Geographers* 77:30–47.

_____. 1993a. *The Colonizer's Model of the World: Geographical Diffusionism and Eurocentric History*. New York: Guilford Press.

_____. 1993b. Mapping the March of History. Paper read at the 1993 annual meeting, Association of American Geographers.

_____. 1994. Robert Brenner in the Tunnel of Time. *Antipode: A Radical Journal of Geography* 26:351–374.

_____. 1999. Environmentalism and Eurocentrism. *Geographical Review*, 89(3): 391–408.

_____, principal author, Frank, A. G., Amin, S., Dodgshon, R., Palan, R., and Taylor, P. 1992. *Fourteen Ninety-Two: The Debate About Colonialism, Eurocentrism, and History*. Trenton: Africa World Press. Originally published in *Political Geography* 11(3), 1992.

_____. Moerman, M., Blaut, R., and Harman, N. 1959. A Study of Cultural Determinants of Soil Erosion and Conservation in the Blue Mountains of Jamaica. *Social and Economic Studies* 8:403–420.

Blench, R. 1993. Ethnographic and Linguistic Evidence for the Prehistory of African Ruminant Livestock, Horses, and Ponies. In Shaw et al., eds., pp. 71–103.

Blumler, M. 1996. Ecology, Evolutionary Theory, and Agricultural Origins. In Harris, ed., pp. 25–50.

Bois, G. 1985. Against the Neo-Malthusian Orthodoxy. In Aston and Philpin, eds. pp. 107–118.

Bray, F. 1984. *Science and Civilization in China: Vol. 6, Part 2. Agriculture* (J. Needham, principal author and editor). Cambridge: Cambridge University Press.

Brenner, R. 1977. The Origins of Capitalist Development: A Critique of Neo-Smithian Marxism. *New Left Review*, no. 104, pp. 25–93.

_____. 1985a. Agrarian Class Structure and Economic Development in Preindustrial England. In Aston and Philpin, eds. 1985, pp. 10–64. Originally published in *Past and Present*, no. 70, 1976.

_____. 1985b. The Agrarian Roots of European Capitalism. In Aston and Philpin, eds. 1985, pp. 213–328. Originally published in *Past and Present*, no. 92, 1982.

_____. 1986. The Social Basis of Economic Development. In J. Roemer, ed., *Analytical Marxism*. Cambridge: Cambridge University Press, pp. 23–53.

Brook, T. 1998. Communications and Commerce. In Twitchett and Mote, eds., pp. 579–696.

Browett, J. 1980. Into the Cul de Sac of the Dependency Paradigm with A. G. Frank. In Peet, ed., pp. 95–112.

Buck, D. 1999. Was It Pluck or Luck That Made the West Grow Rich? *Journal of World History* 10:413–30.

Bulmer, S. 1989. Gardens in the South: Diversity and Change in Prehistoric Maaori Agriculture. In Harris and Hillman, eds., pp. 688–705.

Cain, J., and Hopkins, A. 1986. Gentlemanly Capitalism and British Expansion Overseas: Part I. The Old Colonial System, 1688–1850. *Economic History Review* 39:501–525.

Champakalakshmi, R. 1996. *Trade, Ideology and Urbanization: South India 300 B.C. to A.D. 1300*. New Delhi: Oxford University Press.

Chattopadhyaya, D. 1967. *Lokayata: A Study in Ancient Indian Materialism*, 2nd ed. New Delhi: People's Publishing House.

Cipolla, C. 1965. *Guns, Sails, and Empires: Technological Innovation and the Early Phase of European Expansion, 1400–1700*. New York: Pantheon.

Clark, C. 1977. *Population Growth and Land Use*. London: Macmillan.

Clark, G. 1998. Renting the Revolution. *Journal of Economic History* 58:206–210.

_____. 2000. What Was Real Agricultural Output in England in 1700 Compared to 1850 or 1860? In Mokyr, ed., *The British Industrial Revolution: An Economic Analysis*. Oxford: Westview Press, pp. 206–240.

_____, and Van Der Weif, T. 1998. Work in Progress: The Industrious Revolution. *Journal of Economic History* 58:830–841.

Collins, K., and Roberts, D., eds. 1988. *Capacity for Work in the Tropics*. Cambridge: Cambridge University Press.

Cook, S. W., and Borah, W. 1979. *Essays in Population History*. 3 vols. Berkeley: University of California Press.

Cooper, J. 1985. In Search of Agrarian Capitalism. In Aston and Philpin, eds., pp. 138–191.

Cooper, M. 1980. Town and Country in the Great Transition: Adam Smith et al. on Feudalism and Capitalism. In Peet, ed., pp. 75–82.

Corbridge, S. 1986. *Capitalist World Development: A Critique of Radical Development Geography*. Totowa, NJ: Rowman and Littlefield.

Crone, P. 1989. *Pre-Industrial Societies*. Oxford: Basil Blackwell.

Croot, P., and Parker, D. 1985. Agrarian Class Structure and the Development of Capitalism: France and England Compared. In Aston and Philpin, eds., pp. 79–90.

Crosby, A. W. 1972. *The Columbian Exchange: Biological and Cultural Consequences of 1492*. Westport, CT: Greenwood.

Darby, H. 1952. *The Domesday Geography of Eastern England*. Cambridge: Cambridge University Press.

Davidson, Janet. 1984. *The Prehistory of New Zealand*. Auckland: Longman Pearl.

Diamond, Jared, 1997. *Guns, Germs, and Steel: The Fates of Human Societies*. New York: Norton.

Dirks, N. 1987. *The Hollow Crown: Ethnohistory of an Indian Kingdom*. New York: Cambridge University Press.

Dodgshon, R. 1987. *The European Past: Social Evolution and Spatial Order*. London: Macmillan.

Dutt, R. P. 1943. *The Problem of India*. New York: International Publishers.

Fiedel, S. J. 1987. *Prehistory of the Americas*. Cambridge: Cambridge University Press.

Filesi, T. 1972. *China and Africa in the Middle Ages*. London: Frank Cass.

Flynn, D. 1991. Comparing the Tokugawa Shogunate with Hapsburg Spain: Two Silver-Based Empires in a Global Setting. In Tracy, ed., pp. 332–359.

Flynn, D., and Giraldez, A. 1997. *Metals and Monies in an Emerging Global Economy*. Brookfield, VT: Variorum.

Frank, A. G. 1978. *World Accumulation, 1492–1789*. New York: Monthly Review Press.

———. 1992. Fourteen Ninety-Two Once Again. In Blaut and others, 1992.

———. 1998. *ReORIENT: Global Economy in the Asian Age*. Berkeley: University of California Press.

———, and Gills, B. 1992. The Five Thousand Year World System: An Interdisciplinary Introduction. *Humboldt Journal of Social Relations* 18(1):1–79.

Freund, J. 1968. *The Sociology of Max Weber*. New York: Random House.

Gerth, H., and Mills, C. W. *From Max Weber: Essays in Sociology*. New York: Oxford University Press.

Giblin, J. 1990. Trypanosomiasis Control in African History: An Evaded Issue? *Journal of African History* 31:59–80.

Glover, I., and Higham, C. 1996. New Evidence for Rice Cultivation in South, Southeast, and East Asia. In Harris, ed., pp. 413–441.

Goldstone, J. 1999. Colonizing History: The West Is Best Can't Pass the Test. Paper presented at the Conference on the Origins of the Modern World: Comparative Perspectives from the Edge of the Millennium, University of California at Davis, sponsored by the All-U.C. Group in Economic History, October 15–17.

———. 2000. Review of Landes. *The Wealth and Poverty of Nations*. *Journal of World History* 11:105–111.

Goody, Jack. 1996. *The East in the West*. Cambridge: Cambridge University Press.

Grantham, G. 1989. Agricultural Supply During the Industrial Revolution: French Evidence and European Implications. *Journal of Economic History* 49:43–72.

———. 1993. Divisions of Labour: Agricultural Productivity and Occupational Specialization in Pre-Industrial France. *Economic History Review* 46:478–502.

———. 1999. Contra Ricardo: On the Macro-Economics of Pre-Industrial Economies. *European Review of Economic History* 2:199–232.

Guha, S. 1998. Household Size and Household Structure in Western India c.1700–1950: Beginning an Exploration. *Indian Economic and Social History Review* 35:23–35.

Guyot, A. 1899. *The Earth and Man.* New York: Scribner's, 1899 (original edition, 1849).

Habib, I. 1990. Merchant Communities in Precolonial India. In Tracy, ed., pp. 371–399.

Hall, J. A. 1985. *Powers and Liberties: The Causes and Consequences of the Rise of the West.* Oxford: Basil Blackwell.

_____. 1988. States and Societies: The Miracle in Comparative Perspective. In Baechler et al., eds., pp. 20–38.

_____, and Ikenberry, G. 1989. *The State.* Minneapolis: University of Minnesota Press.

Harlan, J. 1995. *The Living Fields: Our Agricultural Heritage.* Cambridge: Cambridge University Press.

Harris, D., ed. 1996. *The Origins and Spread of Agriculture and Pastoralism in Eurasia.* Washington: Smithsonian Institution Press.

_____, and Hillman, G., eds. 1989. *Foraging and Farming: The Evolution of Plant Exploitation.* London: Unwin Hyman.

Harris, N. 1986. *The End of the Third World.* London: Penguin.

Harrison, P. 1979. The Curse of the Tropics. *New Scientist* 84:602–604.

Hassan, F. 1978. Demographic Archeology. *Advances in Archeological Method and Theory: Volume 1.* ed. M. Schaffer. New York: Columbia University Press.

Heijdra, M. 1998. The Socioeconomic Development of Rural China During the Ming. In Twitchett and Mote, eds., pp. 417–578.

Hilton, R. 1980. Individualism and the English Peasantry. *New Left Review,* no. 120, pp. 109–111.

_____, 1992. *English and French Towns in Feudal Society: A Comparative Study.* Cambridge: Cambridge University Press.

_____, ed. 1976. *The Transition from Feudalism to Capitalism.* London: NLB.

Hodgson, M. 1974. *The Venture of Islam.* 3 vols. Chicago: University of Chicago Press.

_____. 1993. *Rethinking World History,* ed. E. Burke, III. Cambridge: Cambridge University Press, 1993.

Hopkins, A. 1973. *An Economic History of West Africa.* New York: Columbia University Press.

Hoyle, R. 1990. Tenure and the Land Market in Early Modern England: Or a Late Contribution to the Brenner Debate. *Economic History Review* 43:1–20.

Hucker, C. 1998. Ming Government. In Twitchett and Mote, eds., pp. 9–105.

Huntington, E. 1924. *Civilization and Climate,* 3rd ed. New Haven: Yale University Press.

James, C. L. R. 1938. *The Black Jacobins: Toussaint L'Ouverture and the San Domingo Revolution.* London: Secker and Warburg.

Jones, E. L. 1981. *The European Miracle: Environments, Economies, and Geopolitics in the History of Europe and Asia.* Cambridge: Cambridge University Press.

———. 1988. *Growth Recurring: Economic Change in World History*. Oxford: Clarendon Press.

Katz, C. 1993. Karl Marx on the Transition from Feudalism to Capitalism. *Theory and Society* 22:363–390.

Kerridge, E. 1967. *The Agricultural Revolution*. London: Allen and Unwin.

Kertzer, D. 1989. The Joint Family Household Revisited: Demographic Constraints and Household Complexity in the European Past. *Journal of Family History* 14:1–16.

King, M. 1989. *Moriori: A People Discovered*. Auckland: Viking Penguin.

Kosambi, D. D. 1969. *Ancient India: A History of Its Culture and Civilization*. New York: Meridian Books.

Kumar, D. 1985. Private Property in Asia: The Case of Medieval South India. *Comparative Studies in Society and History* 27:340–366.

Kuppuram, G., and Kumudamani, K. 1990. *History of Science and Technology in India*. Delhi: Sundeep Prakashan.

Landes, D. 1969. *The Unbound Prometheus: Technological Change and Industrial Development in Western Europe from 1750 to the Present*. Cambridge: Cambridge University Press.

———. 1983. *Revolution in Time*. Cambridge, MA: Harvard University Press.

———. 1998. *The Wealth and Poverty of Nations: Why Some Are So Rich and Some So Poor*. New York: Norton.

Laslett, P. 1988. The European Family and Early Industrialization. In Baechler et al., eds., pp. 234–242.

Le Roy Ladurie, E. 1985. A Reply to Robert Brenner. In Aston and Philpin, eds., pp. 101–106.

Lee, G. 1987. Comparative Perspectives. In M. Sussman and S. Steinmetz, eds., *Handbook of Marriage and the Family*. New York: Plenum.

Lewis, M., and Wigen, K. 1997. *The Myth of Continents: A Critique of Metageography*. Berkeley and Los Angeles: University of California Press.

Lopez, R., and Miskimin, H. 1961–1962. The Economic Depression of the Renaissance. *Economic History Review* 14:408–426.

Lovejoy, A. 1936. *The Great Chain of Being*. Cambridge, MA: Harvard University Press.

Löwith, K. 1982. *Max Weber and Karl Marx*. London: Allen and Unwin.

Macfarlane, A. 1978. *The Origins of English Individualism*. Oxford: Basil Blackwell.

———. 1986. *Marriage and Love in England: Modes of Reproduction 1300–1840*. Oxford: Basil Blackwell.

———. 1988. The Cradle of Capitalism: The Case of England. In Baechler, et. al., eds., pp. 185–203.

MacNeish, R. S. 1991. *The Origins of Agriculture and Settled Life*. Norman: University of Oklahoma Press.

Mann, M. 1986. *The Sources of Social Power: Vol. 1. A History of Power from the Beginning to A.D. 1760*. Cambridge: Cambridge University Press.

———. 1988. European Development: Approaching a Historical Explanation. In Baechler et al., eds., pp. 6–19.

_____. 1993. *The Sources of Social Power: Vol. 2. The Rise of Classes and Nation-States, 1760–1914.* Cambridge: Cambridge University Press.

Mann, S. 1987. *Local Merchants and the Chinese Bureaucracy, 1750–1950.* Stanford: Stanford University Press.

Marks, R. 1998. *Tigers, Rice, Silk, and Silt: Environment and Economy in Late Imperial South China.* Cambridge: Cambridge University Press.

Marx, K., and Engels, F. 1976. *The German Ideology.* Moscow: Progress Publishers.

McNeill, W. 1976. *Plagues and Peoples.* Garden City, NY: Anchor Books.

Miller, S., Latham, A., and Flynn, D. 1998. *Studies in the Economic History of the Pacific Rim.* London: Routledge.

Minchinton, W. 1969. *The Growth of English Overseas Trade.* London: Methuen.

Mokyr, J., ed. 1999. *The British Industrial Revolution: An Economic Analysis.* Oxford: Westview Press.

Mukerji, S. 1969. *The Republican Trend in Ancient India.* Delhi: Munshiram Manoharlal.

Nag, M. 1980. How Modernization Can Also Increase Fertility. *Current Anthropology* 21:571–588.

Needham, J. 1985. *Gunpowder as the Fourth Estate East and West.* Hong Kong: Hong Kong University Press.

_____, and collaborators. 1954–. *Science and Civilization in China.* Cambridge: Cambridge University Press.

_____, and collaborators. 1962. *Science and Civilization in China: Vol. 4, Part 1. Physics and Physical Technology: Physics.* Cambridge: Cambridge University Press.

_____, and collaborators. 1965. *Science and Civilization in China: Vol. 4, Part 2. Physics and Physical Technology: Mechanical Engineering.* Cambridge: Cambridge University Press.

Nye, P., and Greenland, D. 1960. *The Soil Under Shifting Cultivation.* Farnham Royal, England: Commonwealth Agricultural Bureaux.

O'Brien, P. K. 1996. Path Dependency, Or Why Britain Became an Industrialized and Urbanized Economy Long Before France. *Economic History Review* 49:213–249.

Orwin, C. S., and Orwin, C. S. 1967. *The Open Fields.* Oxford: Clarendon.

Overton, Mark. 1996. *Agricultural Revolution in England.* Cambridge: Cambridge University Press.

Pacey, A. 1976. *The Maze of Ingenuity.* Cambridge, MA: MIT Press.

Parain, C. 1966. The Evolution of Agricultural Technique. In M. Postan and H. J. Habakkuk, eds., *The Cambridge Economic History of Europe.* Vol. 1. *The Agrarian Life of the Middle Ages.* pp. 136–145. Cambridge: Cambridge University Press.

Peet, R., ed. 1980. *An Introduction to Marxist Theories of Development.* Canberra: Research School of Pacific Studies, Australian National University.

_____. 1998. *Modern Geographical Thought.* Oxford: Blackwell.

Perlin, F. 1993. *The Invisible City: Monetary, Administrative and Popular Infrastructures in Asia and Europe, 1500–1900.* Brookfield, VT: Variorum.

Pires, T. 1944. *The Suma Oriental.* London: Hakluyt Society.

Pomeranz, K. 1993. *The Making of a Hinterland: State, Society, and Economy in Inland North China 1853–1937.* Berkeley: University of California Press.

———. 1998. DeLong on David Landes. Electronic text, H-NET for World History. March 26, 1998.

———. 1999. From "Early Modern" to "Modern" and Back Again: Levels, Trends, and Economic Transformation in 18th–19th Century Eurasia. Paper presented at the Conference on the Origins of the Modern World: Comparative Perspectives from the Edge of the Millennium, University of California at Davis, sponsored by the All-U.C. Group in Economic History, October 15–17.

———. 2000. *The Great Divergence: China, Europe, and the Making of the Modern World Economy.* Ann Arbor: University of Michigan Press. In press.

Porter, P., and Sheppard ,E. 1998. *A World of Difference: Society, Nature, Development.* New York: Guilford Press.

Pounds, N. 1990. *An Historical Geography of Europe: 450 B.C.–A.D. 1330.* Cambridge: Cambridge University Press.

Purcell, Victor. 1951. *The Chinese in Southeast Asia.* Oxford: Oxford University Press.

Richards, J. F. 1997. Early Modern India in World History. *Journal of World History* 8:197–210.

Ritter, K. 1865. *Comparative Geography.* Translated by W. Gage. Philadelphia: Lippincott.

Rodinson, M. 1973. *Islam and Capitalism.* New York: Pantheon.

Ross, E. 1998. *The Malthus Factor.* London: Zed books.

Rostow, W. W. 1962. *The Stages of Economic Growth.* Cambridge: Cambridge University Press.

Rowe, W. 1984. *Hankow: Commerce and Society in a Chinese City, 1769–1889.* Stanford: Stanford University Press.

Russell, W. 1967. *Man, Nature and History.* London: Aldus.

Said, E. 1978. *Orientalism.* New York: Random House.

Sharma, R. S. 1966. *Light on Early Indian Society and Economy.* Bombay: Manaktalas.

Shaw, T., et al., eds. 1993. *The Archaeology of Africa.* Routledge.

Simonsen, R. 1944. *História econômica do Brasil, 1500–1820,* 2nd ed. São Paulo: Companhia Editora Nacional.

Sivin, N. 1984. Why the Scientific Revolution Did Not Take Place in China—Or Didn't It? In E. Mendelsohn, ed., *Transformation and Tradition in the Sciences.* Cambridge: Cambridge University Press, pp. 531–554.

———, and Nakayama, S. 1973. *Chinese Science.* Cambridge, MA: MIT Press.

Slicher van Bath, B. H. 1963. *The Agrarian History of Western Europe: A.D. 500–1850.* London.

Smith, C. T. 1969. *An Historical Geography of Western Europe Before 1800.* London: Longman.

Steward, J. 1955. *Theory of Culture Change: The Methodology of Multilinear Evolution*. Urbana: University of Illinois Press.

Stone, L. 1977. *The Family, Sex, and Marriage in England 1500–1800*. New York: Harper & Row.

Subrahmanyam, S., ed. 1990. *Merchants, Markets, and the State in Early Modern India*. Delhi: Oxford University Press.

Subramanian, L. 1999. India's International Economy, 1500–1800. *Indian Historical Review* 25:38–57, 269–288.

Taylor, P. 1989. The World-Systems Project. In R. Johnston and P. Taylor, eds., *A World In Crisis? Geographical Perspectives*, 2nd ed. Oxford: Basil Blackwell.

Taeuber, I. 1970. The Families of Chinese Farmers. In M. Freedman, ed., *Family and Kinship in Chinese Society*. Stanford: Stanford University Press.

Thapar, R. 1982. Ideology and the Interpretation of Early Indian History.*Review* 5:389–412.

Thorndyke, L. 1943. Renaissance or Prenaissance? *Journal of the History of Ideas* 4:65–74.

Titow, J. Z. 1969. *English Rural Society, 1200–1350*. London, Allen & Unwin.

Torras, J. 1980. Class Struggle in Catalonia. *Review* 4:253–265.

Tracy, J., ed. 1991. *The Political Economy of Merchant Empires*. Cambridge: Cambridge University Press, pp. 187–200.

Turshen, M. 1987. Population Growth and the Deterioration of Health: Mainland Tanzania, 1920–1960. In Cordell, D., and Gregory, J., eds., *African Population and Capitalism: Historical Perspectives*. Boulder: Westview Press

Twitchett, D., and Mote, F. eds. 1998. *The Cambridge History of China: Vol. 8. The Ming Dynasty*. Cambridge: Cambridge University Press.

Van Leur, J. C. 1955. *Indonesian Trade and Society*. The Hague: W. van Hoeve.

Venturi, F. 1963. The History of the Concept of Oriental Despotism in Europe. *Journal of the History of Ideas* 24:133–143.

Von Glahn, Richard. 1996. *Fountain of Fortune: Money and Monetary Policy in China, 1000–1700*. Berkeley: University of California Press.

Wallerstein, I. 1980. *The Modern World System, Volume 2*. New York: Academic Press.

Wang Gung Wu. 1991. Merchants Without Empire: The Hokkien Sojourning Communities. In Tracy, ed. (1991).

Watson, A. 1983. *Agricultural Innovation in the Early Islamic World: The Diffusion of Crops and Farming Techniques, 700–1100*. Cambridge: Cambridge University Press.

Webb, W. P. 1951. *The Great Forntier*. Austin: University of Texas Press.

Weber, Marianne. 1975. *Max Weber: A Biography*. New York: Wiley.

Weber, Max. 1951. *The Religion of China*. New York: Free Press, pp. 230–232.

_____. 1958a. *The Protestant Ethic and the Spirit of Capitalism*. New York: Scribners.

_____. 1958b. *The City*. Glencoe: Free Press.

_____. 1968. *Economy and Society*. 2 vols. Berkeley: University of California Press.

_____. 1976. *The Agrarian Sociology of Ancient Civilizations*. London: NLB.

_____. 1981. *General Economic History*. New Brunswick: Transaction Books.

Werner, K. F. 1988. Political and Social Structures of the West, 300–1300. In Baechler et al., eds., pp. 169–184.

White, L., Jr. 1962. *Medieval Technology and Social Change*. London: Oxford University Press.

_____. 1968. The Historical Roots of Our Ecological Crisis. In White 1982, p. 79. Originally published in *Science*, March 10, 1967.

_____. 1982. *Machina Ex Deo: Essays in the Dynamism of Western Culture*. Cambridge, MA: MIT Press.

Williams, E. 1944. *Capitalism and Slavery*. Chapel Hill: University of North Carolina Press.

Wittfogel, K. 1957. *Oriental Despotism*. New Haven: Yale University Press.

Wolf, E. 1982. *Europe and the Peoples Without History*. Berkeley and Los Angeles: University of California Press.

Wong, R. Bin. 1997. *China Transformed: Historical Change and the Limits of European Experience*. Ithaca: Cornell University Press.

_____. 1998. Chinese Views of the Money Supply and Foreign Trade, 1400–1850. In Miller, Latham, and Flynn, eds., pp. 172–180.

_____. 1999. Political Economies of Agrarian and Merchant Empires Compared: Miracles, Myths, Problems, Prospects. Unpublished manuscript.

Wunder, H. 1985. Peasant Organization and Class Conflict in Eastern and Western Germany. In Aston and Philpin, eds., pp. 91–100.

Yang Lien-sheng. 1970. Government Control of Urban Merchants in Traditional China. *Tsinghua Journal of Chinese Studies* 8:186–206.

Young, R. 1990. *White Mythologies: Writing History and the West*. London: Routledge.

Index

About the Author

J. M. BLAUT is Professor of Geography at the University
of Illinois at Chicago. He studies the historical and political geography of
European colonialism and has carried out research in the Caribbean region, South America, and Southeast Asia. Dr. Blaut is the author of four
previous volumes: *The National Question: Decolonizing the Theory of Nationalism*, *Aspectos de la cuestión nacional en Puerto Rico* (with Loida
Figueroa), *Fourteen Ninety-Two: The Debate on Colonialism, Eurocentrism,
and History*, and *The Colonizer's Model of the World: Geographical Diffusionism and Eurocentric History*. Dr. Blaut has also written scholarly pa-
pers on peasant agriculture and on environmental cognition in preschool
children. He studied at the University of Chicago, the Imperial College
of Tropical Agriculture, and Louisiana State University. In 1997 he was
warded Distinguished Scholarship Honors by the Association of America
Georgraphers.